**Kévin LECERF**

**Condominiums in financial difficulty**

Kévin LECERF

# Condominiums in financial difficulty

## Are the prevention and redress mechanisms effective?

ScienciaScripts

**Imprint**
Any brand names and product names mentioned in this book are subject to trademark, brand or patent protection and are trademarks or registered trademarks of their respective holders. The use of brand names, product names, common names, trade names, product descriptions etc. even without a particular marking in this work is in no way to be construed to mean that such names may be regarded as unrestricted in respect of trademark and brand protection legislation and could thus be used by anyone.

Cover image: www.ingimage.com

This book is a translation from the original published under ISBN 978-620-3-41595-7.

Publisher:
Sciencia Scripts
is a trademark of
Dodo Books Indian Ocean Ltd., member of the OmniScriptum S.R.L Publishing group
str. A.Russo 15, of. 61, Chisinau-2068, Republic of Moldova Europe
Printed at: see last page
**ISBN: 978-620-3-67344-9**

**Acknowledgements**

This thesis concludes two enriching years both from a legal and technical point of view, and was an experience rich in instruction. The result of this work of several months would not have been possible without the intervention of certain people whom I particularly want to thank.

Mr. Laurent PROGEAS, my training master, for having welcomed me in his office of surveyor. His ability to explain things, his experience in the syndic environment, as well as his availability were beneficial to me in the progress and structuring of my work. Thank you for your confidence and for the interest you have shown in me.

Mrs. Elisabeth BOTREL who was a very good referent professor since her availability, her extreme reactivity, as well as her wise advice strongly allowed me to establish a clear and complete report on the subject.

All the members of the firm for their more than warm spirit. I also thank them for giving me time for explanations.

Mrs. Alice Lecaudey, in charge of the Habitat mission at Saint-Malo Agglomeration, as well as Corinne Legrand, referent in charge of technical urbanism at Rennes Metropole, for having allowed me to obtain an external vision on the subject. Thank you for having granted me this interview.

Thanks to you.

## List of abbreviations

AFUL : Association Foncière Urbaine Libre
GA : General Assembly
AJDA : Legal news of administrative law (review)
AJDI : Real estate law news (review)
ALUR: Access to housing and renovated urbanism (law)
ANAH : National Agency for Housing
ANGC : National Association of Condominium Managers
ARC : Association of co-ownership managers
ASL : Association Syndicale Libre
CA: Court of Appeal
CAA: Administrative Court of Appeal

CCH : Code de la construction et de l'habitat

CE: Council of State
ELAN: Evolution of housing, development and digital (law)
EPCI: Etablissement Public de Coopération Intercommunale (Public Establishment for Intercommunity Cooperation)
OPAH: Programmed Operation for the Improvement of Housing
PDLHI: Departmental Pole for the Fight against Substandard Housing
PLU : Local Urbanism Plan
POPAC: Operational program for the prevention and support of condominiums

RDI : Real Estate Law Review

Spec: Especially
SRU: Solidarity and Urban Renewal (law)
TA: Administrative Tribunal
TGI : Tribunal de grande instance
TI : Tribunal d'instance
VOC: Monitoring and Observation of Condominiums

## Table of contents

## Introduction

Since the entry into force of the law of July 10, 1965 establishing the status of condominiums, it is however only in the 1990s that the legislator has been particularly interested in reducing the number of condominiums in financial difficulty.

*"Noting that the law of collective procedures was clearly unsuited to the particular situation of trade unions, the drafters decided to create within the law of 1965 a section specifically devoted to co-ownerships in difficulty "*[1]. Consequently, the law relating to housing of July 21, 1994 included in the law of 1965 the expression of

"This article specifies that a co-ownership is considered to be in difficulty "if the financial equilibrium of the syndicate of co-owners is seriously compromised or if the syndicate is unable to provide for the property. This article specifies that a co-ownership is considered to be in difficulty "*if the financial equilibrium of the syndicate of co-owners is seriously compromised or if the syndicate is unable to provide for the conservation of the immovable*". This desire to identify co-ownerships in difficulty is accompanied by a support body for them since a provisional administrator will be appointed to restore normal operation. "*This designation is different from that to which private companies are subjected, but it is inspired by it "*[2]. Through the definition mentioned, the legislator wished to remain broad to let a maximum of co-ownerships benefit from this assistance. However, in spite of this will, it must be noted that this measure has not been able to completely stop this phenomenon since to date the ANAH has identified 18% of co-ownerships showing at least signs of fragility3. Aware of the large number of condominiums in difficulty, the legislator has multiplied the number of measures aimed at correcting them. To date, no less than 26 articles are devoted to co-ownerships in difficulty in the law of 1965. This multiplication of measures has appeared progressively. Indeed, whether it is the SRU law in 2000, the MOLLE law in 2009, the ALUR law in 2014, the CAP law in 2017, or the ELAN law in 2018, all of them include provisions dedicated to curbing the number of these failing condominiums4.

---

[1] GUILHEM Gil, "Copropriétés en difficulté - Garanties de recouvrement et syndicats de copropriétaires en difficulté", *JurisClasseur Copropriété*, Fasc.84, February 2018, spec. n°41

[2] SAINT ALARY HOUIN Corinne, " Les copropriétés en difficulté ", *Revue Droit et Ville* n°52, 2001, p 103

[3] ANAH, "Plan Initiative copropriétés Des dispositifs pour accompagner les interventions locales", 2019

4 In addition, an amount of two billion five hundred thousand euros has been retained in 2018 by the government through the "Co-ownership Initiative Plan" to initiate a policy of improving the housing stock in co-ownership. This plan of unprecedented scope demonstrates that the fight against condominiums in difficulty has become an emergency.

This accumulation of measures reinforces the idea that It is difficult to stop this phenomenon and leads us to ask whether the measures put in place are really effective?

In order to ensure a serious policy of fight against co-ownerships in financial difficulty, it is naturally advisable to first establish a diagnosis of the potential origins of the occurrence of a difficulty in order to then avoid them. "*The origin of a co-ownership in difficulty can be diverse. Social (overpopulation, precariousness, over-indebtedness, high proportion of non-occupying co-owners...), technical (defective construction,dilapidation, insalubrity...), legal (negligence of the managing agent, absence of co-ownership rules...), accidental* " [5]. Following the BRAYE report in 2012, advocating preventive rather than curative action, the emergence of devices aimed at fighting against these causes appeared. Only the devices aiming to avoid legal and then social causes will be mentioned since they are recurrent causes6.

Concerning the legal causes, the recrudescence of litigation on the payment of an amount7 , which then provokes a financial difficulty, raises questions on the relevance of the drafting of the co-ownership regulations as well as of the management related to them. "*The CRA never ceases to remind us that a system of governance and management jeopardized by financial imbalances and bodies with unbalanced powers facilitates the arrival of financial difficulties* "[8]. This problem which may prove to be a source of financial difficulties therefore raises questions as to the possibilities of the drafters of the co-ownership regulations (who may be surveyors) to draft relevant regulations in order to avoid any litigation, in particular concerning the payment of an amount. On the other hand, in spite of a relevant drafting, if it turns out that the management of the co-ownership by the syndic is doubtful, then the litigations seen previously cannot be avoided. In fact, "*various studies9 report multiple causes which may be due to the poor management of the syndic*".

---

[5] ROUX Jean-Marc, Prévenir et redresser les copropriétés en difficulté, Edilaix, Point de droit, p.250, 2016

[6] LEBEL Christine, "Procedures applicable to syndicates of co-owners in difficulty", *Revue des procédures collectives*, n°5, September 2014

[7] According to a report by Pierre Capoulade, Caroline Moreau and Brigitte Munoz Perez, unpaid claims account for more than two-thirds of legal actions

[8]ARC-copro.fr, 1 year after launch1storcod, feedbackexperience,2017 ;https://arc-copro.fr/documentation/1-after-launching-the-1st-orcod-of-france-returns-of-experience-of-larc-clichy, accessed 04/01/2020

[9] See, in particular, the report on the hearing held by the Economic and Social Council on December 10, 2001 on condominiums in difficulty, which can be consulted on the i.vill.gouv.fr website.

[10]. All of these elements also raise questions about the effectiveness of the measures aimed at regulating the status of trustee.

Concerning the social causes, "*among the difficulties faced by trustees in the management of a condominium is often the impossibility of having the general assembly vote on certain decisions despite their appropriateness or necessity* "[11]. This difficulty, which can indirectly become financial, raises questions about the involvement of the co-owners in their building since if, at the time of the purchase, they are not aware of all the impacts that this generates (weight of the charges, obligation to participate in the general assemblies...), this can quickly influence the financial health of the co-ownership.

Although the elimination of the causes seen previously is an important axis to ensure a serious policy of fight against co-ownerships in financial difficulty, this axis, which can be likened to the prevention of difficulties, seems to be able to be reinforced by devices aiming at identifying and then accompanying the potentially fragile co-ownerships.

In addition to all these questions, it is also necessary to question the effectiveness of the measures allowing for the recovery of a situation of difficulty. The surveyor acting as a trustee12 is a major actor in the recovery process and must therefore have a perfect knowledge of the proposed mechanisms. In order to re-establish a normal situation, some authors, such as Jean-Marc ROUX, point out that the measures taken against co-ownerships in difficulty must be adapted to the degree of difficulty encountered by the co-ownership. Thus, three levels regularly emerge from the writings of academics and legal practitioners: fragile co-ownerships, co-ownerships in difficulty and degraded co-ownerships. Although these levels are not explicitly included in the 1965 law, some of its articles can be associated with them13.

---

[10] REGNAUT-MOUTIER, "Le traitement des difficultés des syndicats de copropriétaires", *Revue des procédures collectives*, n°3, May 2010

[11] ROUX Jean-Marc, " Réflexions sur l'assouplissement des conditions d'adoption des décisions en assemblée générale de copropriété ", *Loyers et copropriété,* n°6, Juin 2007

[12] Article 8-1 Law of 1965: Membership of the order is not incompatible with the exercise of an activity of real estate mediation

[13] The mandatary ad hoc (Article 29-1 A) for fragile co-ownership. The provisional administrator (article 29-1) for co-ownership in difficulty. The reinforced provisional administrator (Article 29-11) for deteriorated co-ownership.

This distinction is necessary to undertake a judicious recovery of the normal functioning of the co-ownership since the different levels allow to apprehend the needs of the co-ownership.

First of all, a fragile co-ownership is considered to be "a *co-ownership showing signs of fragility, at risk of entering into a spiral of de-qualification which could eventually lead to difficulties* "[14] . This definition therefore requires us to examine the different signs of fragility in order to study the effectiveness of possible recovery measures.

The level of co-ownership in difficulty remains defined as seen previously in article 29-1 of the law of July 10, 1965. "*The opening criterion is no longer based on a threshold level of unpaid bills in relation to the voted budget, but results from a general situation* "[15]. Faced with such a situation, it is necessary to compensate for the lack of cash flow generally due to the need for heavy work. Such a level of difficulty thus leads the co-ownership to ask itself: how could it find financial resources to enable it to recover?

The ultimate stage, which is the degraded co-ownership, is defined as "*a co-ownership which presents risks for the safety of the co-owners and for which the conservation of the building seems complicated* "[16]. Contrary to the previous stage, we understand here that it is a general situation irremediably compromised. Faced with such a situation having an urban consequence, it is interesting to question the implication of the communities on the degradation of the private housing stock.

Through all the questions raised, 2 major axes stand out, they are the prevention of the appearance of financial difficulties (I) and the recovery of co-ownerships in financial difficulty (II).

---

[14] DEVRY Pascal, " Le point sur : The new ANAH aid for fragile condominiums", *cahier des copropriétés en difficulté,* n°11, p.2, 2017

[15] BRESSON Réné, TULIER-POLGE Florence, MENJUCQ Michel, " Les copropriétés en difficulté-Table ronde ", *Revue des procédures collectives,* n°1, 2017

[16] ROUX Jean-Marc, Prévenir et redresser les copropriétés en difficulté, Edilaix, Point de droit, p.250, 2016

# I The prevention of the appearance of financial complications in co-ownership, a late awareness by the legislator

Following the Braye report in 2012, the legislator moved towards the implementation of measures to prevent financial difficulties in condominiums. Indeed, "*focusing public action solely on the most dramatic cases would be a serious mistake and the absence of an overall policy on condominiums represents the assurance of seeing other areas of difficulty develop* "[17]. This is why, in order to "*avoid piecemeal actions* "[18] , mechanisms allowing to avoid the frequent causes of financial difficulties have progressively been put in place and must be well used. (I.1) In addition, the mechanisms for reporting on co-ownerships showing signs of fragility are effective, but still too little used. (I.2)

## I.1 The multiplication of mechanisms to fight against the recurrent causes of financial distress: a major initiative

Many recent measures aimed at alleviating the causes of financial difficulties have been introduced and are effective if they are mastered. First of all, ensuring a sound structure19 of the co-ownership makes it possible to avoid any dispute indirectly causing unpaid bills (I.1.1). To do this, it is possible to set up special charges, to divide the syndicate if necessary and to appoint a competent syndic. Furthermore, in order to reduce another indirect cause of financial difficulties, the legislator wished to remedy the poor involvement of the co-owners. (I.1.2) By envisaging the possibility of anticipating future expenses20 , by facilitating access to general meetings21 , or by reducing the number of indelicate co-owners. These measures are certainly effective in preventing financial difficulties, but they are potentially difficult to implement.

---

[17] BRAYE Dominique, "Preventing and curing difficulties in condominiums. A priority for housing policies", 2012

[18] BRAYE Dominique, Precity

[19] Judicious drafting of the co-ownership by-laws and sound management of the syndic

[20] By the possibility of feeding an extranet and a work fund, by the possibility of carrying out a DTG

[21] By authorizing participation in the meeting by electronic means, by authorizing voting by form or by proxy

### I.1.1 A relevant structuring of the co-ownerships for a healthy operation

In ten years, litigation related to the payment of charges which ultimately leads to financial difficulty has increased by twenty-nine percent and represents two thirds of the litigation in condominiums22. Although it is possible to take out insurance against unpaid charges to avoid financial difficulty, it is wiser to prevent these disputes. Many of these disputes could be anticipated by a " healthy " life in co-ownership. To obtain this, it is necessary to judiciously distribute the co-ownership expenses and to ensure an optimal management. The legislator has therefore foreseen, through special charges, the possibility of judiciously distributing the charges, thus allowing the adoption of a logic of equity and acceptability (I.1.1.1). In the same perspective, it is possible to divide the syndicate even if this recourse is expensive, complex and not always possible to implement (I.1.1.2). In addition to this judicious drafting, it is necessary to appoint a competent syndic to manage the co-ownership. Competences which the voluntary trustees do not necessarily have since they are not obliged to be trained (I.1.1.3).

#### I.1.1.1 Anticipation of unpaid bills thanks to special charges

To ensure consistency in the distribution of charges, it is possible to establish special charges. Often associated with special common areas23 (or common areas with exclusive use)[24] , "*this mechanism allows the co-owners to feel that they are only responsible for the expenses related to the areas they actually use* "[25] . This makes it possible to avoid unpaid bills due to co-owners who refuse to pay because they do not feel concerned by the measures taken. Aware that this mechanism makes it possible to prevent certain causes of unpaid bills, the legislator, through the ELAN law, has come to define26 and reinforce the management of these special common areas.

---

[22] GUEGAN-GELINET Laurence, " Le contentieux de recouvrement des charges de copropriété après les lois ALUR et ELAN ", *Transversale immobilière,* pp.12-18, n°142, mai-juin 2019

[23] CA Paris, 23rd ch. B, Nov. 25, 1999, commented by GIVERDON Claude, RDI 2000 p.89: "*to be able to specialize charges, it is not for all that indispensable to formalize the creation of special common parts*"

[24] CA Poitiers, June 15, 2011, JurisData n° 2011-031768 commented by VIGNERON Guy, " Parties communes : droit d'usage exclusif et répartition des charges ", *Loyers et Copropriété,* n° 4, April 2012, comm. 126

[25] DEBESSELLE Clémence, *La spécialisation des charges et des parties communes en copropriété ",*
Geometric and Topographic Engineer thesis, ESGT, 2016, p.67

[26] Article 6-2 of the law of July 10, 1965

It is now possible to meet in a special general assembly for decisions concerning these parts only, "*this is a novelty which could prove useful especially in large complexes to avoid having to form secondary syndicates* "[27] . However, "*the drafter of the SDA and the condominium by-laws must use the specialization of the common portions in a thoughtful manner, in order to avoid conflicts* "[28] . Indeed, this specialization must not be the object of too great a complexity involving a lack of understanding of the different undivided rights. All the more so since the relationship or otherwise between special charges and special common portions is not always understandable by all.

"*Specialization remains more appropriate for large condominiums where the initial management is already more complex. This tool will lighten and clarify the situation; but it is not to be preferred for small co-ownerships of a few lots where it would have the opposite effect, multiplying property rights, quotas, expenses and meetings where this would not be necessary29* ". Moreover, in certain complex situations, it is difficult to envisage a coherent distribution of expenses without resorting to the division of the syndicate.

### I.1.1.2 Anticipation of unpaid bills by dividing the syndicate

In the case of large condominiums with several independent buildings, the adoption of such a division of the syndicate which aims at "*simplifying and making each entity autonomous* "[30] is sometimes necessary to ensure a sound management. Indeed, "*the lack of power of each co-owner, his feeling of not being able to influence decisions generally leads to discouragement and absenteeism* "[31] . Moreover, this can facilitate the emergence of co-owners in bad faith who refuse to pay their charges, because they do not feel concerned by the measures undertaken on the other buildings32. Faced with such a situation, despite the certain advantages of resorting to this type of However, the conditions imposed and the costs involved restrict a certain number of co-ownerships from applying this approach.

---

[27] BOHBOT Charles, Loi ELAN : parties communes spéciales et modification du règlement, bjavocat.com, 2019

[28] DEBESSELLE Clémence, *La spécialisation des charges et des parties communes en copropriété,* Mémoire d'ingénieur Géomètre et Topographe, ESGT, p.67, 2016

[29] DEBESSELLE Clémence, supra

30DALBIN Hugo, *Does the application of the technique of the division in volumes meet abusive uses?* ,

Master's thesis Identification, Land Use Planning and Management, ESGT, p.60, 2019

[31] SOCAF, Fact *Sheet,* #793, July-August 2018

[32] SOCAF, supra

Indeed, it should be remembered that the division of the syndicate "*entails a breakdown of the expenses which until then had been distributed among all the lots* "[33] , which can help prevent conflicts of interest. It may be done through the establishment of one or more secondary syndicates (in which case the main syndicate will remain in place) or through a division of co-ownership (consisting of the removal of one or more buildings from the initial syndicate to constitute a separate property)[34]. The provisions of section 27 of the 1965 Act impose two conditions for the constitution of a secondary syndicate. "*Firstly, there must be a plurality of buildings, and a special assembly must be held, taking its decision under the majority conditions provided for in article 25.*[35] The constitution of a secondary syndicate is not a matter of a single building, but of a single *property.* The constitution of a secondary syndicate is therefore not too cumbersome a procedure to put in place, but will generally be ineffective in preventing unpaid bills in the event of abuse36. Concerning demerger, article 28 of the law of 1965 distinguishes between simple demerger and demerger in volumes. In order to set up a simple division, several independent buildings and a possible division of the ownership of the land are required. In some condominiums, the condition of divisibility of the land cannot be obtained because of complex interlocking, which is why the legislator, with the ALUR law, has created the "volume split "[37]. The legislator also sets conditions for the division into volumes to be possible. There must either be several distinct buildings on a slab, or several homogeneous entities assigned to different uses. It is also necessary that this division be adopted by the General Meeting by an absolute majority38 and that the legal, material and financial conditions be defined. All these conditions are necessary to avoid abusive recourse to such an approach, but this reduces the effectiveness of the system since it strongly restricts access to.

---

[33] TOMASIN Daniel, "La Copropriété", *DALLOZ Action*, chapter 325: Syndicat en difficulté, 2018/2019

[34] Article 28 of the 1965 law

[35] TOMASIN Daniel, supra, chapter 324: Division of the syndicate-Administration-Property

[36] *"the abusive multiplication of secondary unions could only undermine the proper functioning of the co-ownership*; PERINET-MARQUET Hugues, "Accès au logement et urbanisme rénové Loi ALUR du 24 mars 2014", *La semaine juridique,* n°15, avril 2014

[37] Procedure for dividing a co-ownership into volumes

[38] The ELAN Act has made it easier to amend the bylaws by reducing the majority required within 3 years of their adoption. The amendment of the bylaws, which in principle must be voted according to the unanimity rule, can be voted by a simple majority of those present or represented.

This measure.39 Moreover, despite the benefits of these measures, many people question their effectiveness. It has been said, for example, that these remedies are a "*not very effective palliative for the ills suffered by large housing estates* "[40]. This doubt is permitted, because the search for this objective faces many difficulties. The use of this division implies, most of the time, a considerable reworking of the documents "structuring" the co-ownership (the co-ownership regulations, the descriptive state of division...). This reorganization necessarily involves costs that can quickly become significant. As an example, the splitting of a co-ownership of two buildings in Saint-Malo cost 5000€. In more complex cases, the cost can rise very quickly, as in the case of the splitting of a condominium in La Noue, where the operation cost several million euros41. It should be noted that the length of the procedure can also be a brake on the adoption of such an approach by the co-owners. According to Bruno DHONT42 , volume splitting is complex and will only be used in a judicial context. "*However, given the complexity of the management of certain co-ownerships in difficulty, the use of volume splitting offers an effective alternative for a possible recovery* "[43] . If, however, the writer of the co-ownership regulations has been able to ensure a coherent distribution of charges, in order to ensure a healthy structure, favourable to the good economic health of the co-ownership, it remains to ensure that the related management is also mastered.

### I.1.1.3 Anticipation of unpaid bills by a competent syndic

After having made sure that the rules of co-ownership are judiciously drafted, it is advisable to make sure that the appointed syndic is competent. Indeed, the good economic health of a co-ownership strongly depends on the management of the co-ownership since he will have to manage the budget, ensure the recovery of unpaid bills, anticipate the future charges...

---

[39] See DALBIN Pauline, *Scission d'un grand ensemble initialement sous le régime de la copropriété en une division en volumes,* Mémoire de Master Identification, Aménagement et Gestion du Foncier, ESGT, p.49, 2015

[40] 73rd Congress of French Notaries, "Pratique et évolution de la copropriété", Strasbourg, 1976, p.445

[41] Servirlepublic.fr and BERGOZ Edouard, *Les scissions de copropriété comme outil de résolution de situations complexes,* Mémoire d'ingénieur Géomètre et Topographe, ESGT, p.71, 2017

[42] Director of the ARC. According to FULCHERI Elisabeth, *La scission en volumes, une solution pour les copropriétés dégradées ?* Engineering thesis, ESGT, p.102, 2014

[43] DALBIN Hugo, *Does the application of the technique of division into volumes encounter abusive uses?*

Master's thesis Identification, Planning and Land Management, ESGT, p.60, 2019

Aware of this, the *The* legislator has attempted to regulate the status of trustee through obligations. This is a good initiative, but it is necessary to complete it in order to completely stop this potential cause of financial difficulties, whose origin is indirect. Indeed, "*the immobility of the syndicate of co-owners or the lack of training of voluntary trustees often accentuates the fragility of these co-ownerships and leads them, in the long run, into a process of de-qualification* "[44].

First of all, the mission of the syndic must be carried out in a rigorous way at the risk of seeing the state of the co-ownership deteriorate. To avoid any negligence on the part of the syndic, article 18 of the 1965 law as well as article 1992 of the civil code specify that their responsibility can be engaged in case of failure *"for example in the case of a management error or a failure to comply with the duty to advise"* [45]. However, in such a situation, as long as the responsibility of the lax syndic is engaged, the co-ownership continues to be badly managed and to deteriorate. In fact, after sending a formal notice by registered letter with acknowledgement of receipt to the syndic, if the mediation attempt remains unsuccessful, it will be appropriate to change the syndic. This change can be made in several ways. The ideal solution is not to renew the contract, but this requires waiting until the end of the syndic's mandate46 . To avoid such a wait, it is possible, even if "*the law of 10 July 1965 and the decree of 17 March 1967 are relatively discreet on this hypothesis* "47 , to call an exceptional general meeting48 to dismiss the trustee49. Moreover, in order to dismiss the trustee, a majority of the votes of all the co-owners must be expressed in this sense50 . If this is not the case, it will be necessary to take legal action before the judicial court which will designate a provisional administrator pending the appointment of a new trustee.

---

[44] ANAH AND SAINT MALO agglomeration, "Agreement for the implementation of an Operational Program for Prevention and Support of Condominiums (POPAC) in Saint-Malo Agglomeration", 2020-2022, p.3

[45] DOUDOUX Angeline, "Role and responsibility of the trustee in the management of degraded co-ownerships",

*Edition Francis Lefebvre*, March 2019

[46] Article 21 of the 1965 law: a trustee must be put out to tender every 3 years, unless the General Meeting decides otherwise

[47] LAFOND Jacques, "Organization and operation of co-ownership", *JurisClasseur Notarial Formulaire,* Fasc. 60, September 2017, speculative n°13

[48] This must be requested by the trade union council or the co-owners representing at least one quarter of all the co-owners from the syndic. If no reply is received within eight days, the president of the general council may convene the general meeting.

[49] The mandate of a co-ownership trustee may be revoked at any time, *despite the term of the contract (Cass. Civ., April 27, 1988)*

[50] Without a majority, it is also possible to have recourse to a second assembly to hope to obtain the "catch-up majority", this time, it is only the majority of those present and represented which is necessary for the revocation of the trustee

However, dismissal is only possible if there are valid and proven reasons51 which are sometimes difficult to prove. The possibility of revoking a lax trustee is therefore difficult, which can quickly lead to financial difficulties.

However, imperfect management by trustees is not an isolated situation. In fact, despite the many skills that managing agents must have52 , according to the Association des Responsables de Copropriétés, about 15% of co-ownerships are managed by a voluntary managing agent, compared to 5% 10 years ago53. "*The use of a voluntary trustee comes for the most part from a desire to make savings, to involve the co-owners more or to manage current affairs more directly* "[54]. Even if "*the responsibility of the voluntary trustee is often assessed less severely because of the voluntary nature of his mandate* "[55] , the fact remains that he has the same obligations in terms of his role. Unlike professional trustees, they are not required to have professional liability insurance, a management card and a financial guarantee56 . He must only, since the ALUR law in 2014, be part of the co-ownership57 and have civil liability insurance. Although it is impossible to generalize, it appears from the register of co-ownerships that co-ownerships showing signs of fragility are mostly managed by a volunteer syndic. Without specific training, it seems complicated to manage a large condominium alone because of the multitude of knowledge that it requires (on the technical, legal and accounting levels in particular).

---

[51] Cass. 3e civ., Apr. 27, 1988, No. 86-11.718: Bull. civ. III, No. 80; commented by LAFOND Jacques, supra: the trustee can only be dismissed for legitimate reasons

[52] In accordance with the law of 1965, the missions of the syndic are numerous. He must have technical, accounting and legal knowledge in order to fulfil this function which is not at all obvious.

[53] Association des Responsables de Copropriété, "Les syndics bénévoles en copropriété", *Vuibert*, p.252, 2015

[54] TURENNE Paul, "Le syndic bénévole en copropriété", *Informations Rapides de la Copropriété*, n°584, December 2012

[55] MAINAULT Julie, " Syndic bénévole : bien gérer sa copropriété ", *Informations Rapides de la Copropriété*, n°644, December 2018 ; Article 1992 civil code ; The late collection of common charges is considered insufficiently serious to justify the liability of a voluntary syndic V. Paris, 16th Ch., April 24, 1975 : J.C.P. 75, II, 18097, note Guillot).

[56] Article 3 law n° 70-9 of January 2, 1970

[57] Article 17-2 of the law of 1965

"*Apart from the psychological aspect which justifies that a victim hesitates to act against a voluntary trustee* "[58], the major correlation59 between a voluntary trustee and a co-ownership in difficulty comes from the absence of a financial guarantee60 and especially from a lack of training. It is not a question of criticizing voluntary trustees (since this may prove useful for certain co-ownerships), but rather of supervising the conditions of access to this status61 . For example, it could be a matter of imposing on them a compulsory information day62 , even if dematerialized, on the exercise of their function. Without this information day, they would not be able to exercise their mandate or take out insurance. Attendance at this day would allow them to obtain a certificate that would be annexed to the register of the co-ownership. It could also be an evaluation of their capacity to govern for a short period. This measure would allow the professional trustees conducting this information day to find an additional source of revenue while allowing the volunteer trustees to be made aware of their function. In the same vein, concerning training, in fragile areas, it may be interesting to impose the use of a certified trustee "qualiSr Syndic Prévention Redressement "[63] . Having received a validation of skills, specific to co-ownerships in difficulty, by a certified organization, he would be in a better position to redress the situation before it deteriorates.

In spite of certain obstacles to the recourse to the division of the syndicate and the lack of training of voluntary trustees, the measures put in place by the legislator aiming to efficiently structure the co-ownership are globally efficient. However, clumsiness when drafting the co-ownership regulations and the control of the rigour of the trustees are not the only causes of financial difficulties. To ensure a " healthy life ", a vector of good economic health of the co-ownership, it is also necessary to ensure a deep involvement of the co-owners.

---

[58] COUTANT-LAPALUS Christelle, "Some thoughts on civil liability and the syndic of co-ownership", *Loyers et Copropriété n°1,* January 2014, study 1, spec. n°9

[59] Management in its favor can also be a source of hardship, but it is not the primary cause.

[60] The funds held by the voluntary trustee could disappear without any real recourse from the co-owners

[61] This framework should also apply to cooperative trustees

[62] A single day per year seems enough to allow them to master the main aspects of their missions. Moreover, this obligation should not have too heavy financial consequences for the co-ownership.

[63] These are trustees specifically involved in the prevention and recovery of co-ownerships in difficulty.

### I.1.2 Need for a deep involvement of the co-owners

The co-owners, because of their status, are of course essential actors to allow a co-ownership to be healthy. Becoming a co-owner, by purchasing a lot, does not have the same constraints as the acquisition of a building in single ownership. However, this forgetfulness is an indirect factor of financial difficulties since a bad anticipation of future expenses (I.1.2.1), an absence during the general assemblies (I.1.2.2), a negligence in the maintenance of the private lot (I.1.2.3), will have an impact on the whole of the co-ownership. In order to mitigate these indirect causes of financial difficulties due to the co-owners, the legislator has multiplied the number of measures allowing to reduce the number of defaulting co-ownerships. The devices in question are for the anticipation of the charges, the Global Technical Diagnosis and the work plan. In order to facilitate the presence at the general assembly, the legislator has opened the possibility to attend by videoconference, to give power or to vote by mail. Concerning the negligence in the maintenance of the private lot, only devices aiming at identifying and then sanctioning the indelicate co-owners have been implemented and are still insufficient since the trustees encounter difficulties to know the state of maintenance of the private parts.

#### I.1.2.1 The development of information and anticipation methods for future expenses: a necessity

As indicated, the legislator has developed procedures for providing information and anticipating future charges. First of all, it is up to the notary to remind the purchaser of his rights and obligations64 . The notary must also inform the purchaser, thanks to the various information he has been able to gather65 , on the state of the co-ownership, in particular on the last resolutions voted by the general assembly, on the state of the procedures in progress... However, this information stops at the provisional budget which may prove to be erroneous in certain cases, in particular when unforeseen works are carried out. Paying for unforeseen work has always been a real headache for co-ownerships, especially for those who are already carrying heavy loads.

---

[64] It is as soon as the promise is signed that information must be given on the situation within the co-ownership; FREMEAUX Eliane, "The role of the notary in the life of the co-ownership", *Informations rapides de la copropriété,* n°582, 2012; Article 1240 Civil Code; Article 20 law of 1965

[65] Thanks to the dated statement provided by the trustee, for example

"*Especially since the deterioration and technical obsolescence can make very heavy work indispensable* "[66]. To overcome these difficulties, which can indirectly cause financial difficulties, it is possible to multiply the methods of information and, since 2015, they must be gathered within an Extranet which will allow, if it is well supplied, to prevent financial difficulties. In addition, the setting up of a work fund is also effective to allow co-owners to partially anticipate future works.

Provided for in article 18 of the law of 10 July 1965, the Extranet "*is a secure online platform whose aim is to group together the important documents relating to the co-ownership in a dematerialized way, in order to facilitate its access to all* "[67]. The idea is therefore good, but the ARC has noted that it was rarely fed by the syndics. To remedy this, the regulatory power has imposed that certain documents be integrated into it68. It is planned to include documents relating to the management of the building (condominium rules, building maintenance booklet, technical diagnoses, insurance contracts, etc.) as well as documents specific to each condominium owner (individual account of the condominium owner, amount of charges, amount of the share of the work fund, etc.). A section reserved for the trade-union council is also included (general balances of accounts, legal summonses...). Linked to the extranet, there is also a digital logbook69 which should make it possible to know the state of the dwelling and the building and to accompany the improvement of energy performance. Despite the relevance of imposing the integration of the above-mentioned elements, this extranet can be double-edged since the law governing co-ownership only requires an annual update. The situation of a co-owner at a given moment may therefore prove to be erroneous. Moreover, since there can be several General Meetings in one year, the syndic must be vigilant on the feeding of this extranet to avoid any misunderstanding. Moreover, it is regrettable not to allow the notary to have access to this system, even partially, because it would allow to inform more easily the purchaser on the weight of the charges to come, thus indirectly preventing unpaid bills.

---

[66] Dominique Braye, Prévenir et guérir les difficultés des copropriétés. A priority for housing policies, 2012

[67] https://www.syndic-one.com/loi-elan-quels-documents-sur-lextranet/, accessed on 10/02/2020

[68] Decree No. 2019-502 of May 23, 2019 on the minimum list of dematerialized documents concerning co-ownership accessible on a secure online space

[69] Article 182 of the ELAN Act, Article 11 of Act No. 2015-992 of August 17, 2015 on the energy transition for green growth

The main element allowing the weight of future charges to be assessed, which can be fed into the extranet, is the Global Technical Diagnosis70 . This document requires a certain investment, but is essential for the efficient "management" of the co-ownership. The duty of the surveyor to advise the co-owner must not be neglected when choosing a possible implementation of this system. Indeed, this document deserves to exist even before the co-ownership shows signs of fragility or difficulty. Moreover, it is supposed to allow "*to save money by anticipating the works. The purpose is not to end up with a simple detailed list of interventions. This document must propose and evaluate solutions with regard to their cost and degree of relevance, urgency and obligation*. However, at present, it is a document that is not required for all condominiums. It is only compulsory for buildings built more than 10 years ago and subject to co-ownership72 . It is also mandatory for co-ownerships that have been the object of a procedure for insalubrity73 , for buildings that are in danger of collapse74 and for dangerous buildings75. Failing that, the administration can have it carried out ex officio. In his 2012 report, Mr. BRAYE wished to extend this measure to all condominiums. It is indeed a measure that would indirectly reduce the number of co-ownerships in financial difficulty. In addition, this document must show a summary evaluation of the cost and a list of the works necessary for the conservation of the building, by specifying in particular those which should be carried out in the next ten years. This multi-year work plan is a major step forward in the fight against co-ownerships in difficulty since its objective is to anticipate and spread out future expenses. In the absence of the adoption of a DTG, the Elan law wished to impose the establishment of a 10-year work plan for buildings older than 15 years. However, this is a measure that was withdrawn at the last minute, as the Council of State would have issued a negative opinion, which has not been made public. According to Mr. Torrollion, president of the Fédération Nationale de l'Immobilier, "*It is distressing: the flagship measure of this.*

---

[70] Article L 731-1 CCH

[71] BADUEL Yves, "Fatality? Lack of technical mastery? ", *Geomètre,* n°2170, June 2019 p.28

[72] Article L 731-4 CCH

[73] Article L 1331-26 code de la santé publique

[74] Article L 511-1 CCH

[75] Article L 129.1 CCH

The *ordinance does not even appear in it even though it had been unanimously approved by all the actors of the profession* "[76]. It is indeed a measure that could have facilitated the anticipation of future expenses, thus reducing the number of condominiums in financial difficulty, but the duration of 10 years could have had the opposite effect by imposing work "*that it would have been more coherent to spread over a longer period* "[77].

In addition to these (extranet, digital logbook, DTG), article 58 of the ALUR law now makes it compulsory, through article 14-2 of the law of 1965, to set up a works fund for co-ownerships of a certain size. This fund "*is intended to meet the expenses resulting from the work prescribed by the laws and regulations and the work decided by the general assembly of co-owners not included in the provisional budget* "[78] . It therefore appears to be a preventive measure for co-ownerships in difficulty, since it aims to anticipate the payment of subsequent works. Even if this work fund is not imposed on all co-ownerships, the exceptions remain very limited79 , thus increasing its effectiveness. "*The sums paid (contributions) are attached to the lot and definitively acquired by the syndicate. They are therefore legally expenses* "[80]. The amount of the contribution is greater than or equal to five percent of the annual budget. This percentage is coherent but should be relativized according to the size of the condominium, since this percentage will allow for a fairly substantial budget for large condominiums but not for small condominiums. Moreover, it would be a pity if this fund itself led to the appearance of co-ownerships in difficulty due to a sum requested that is too large. This same article 14-2 of the 1965 law also provides for the end of the obligation to pay into the works fund when the fund reaches an amount higher than the provisional budget. The compulsory nature of this measure makes it possible to recover.

---

[76] Batinfo.fr, 31/10/2019, accessed 12/12/2019, https://batinfo.com/actualite/la-reforme-des-coproprietes- valid-less-measures-central-obligation-plan-of-work_14101

[77] ARC-copro.fr, "Un plan pluriannuel de travaux retoqué logiquement par le Conseil d'Etat", 19/11/2019 [78] BUCHER Charles-Édouard, "COPROPRIÉTÉ-Copropriétaires. - Duties of co-owners. "*JurisClasseur Civil Code,* Fasc. 31-1, January 2020, spec. no. 11

[79] All co-ownerships are concerned by this obligation, except for 3 cases: Buildings less than five years old; buildings with less than ten lots for which a unanimous vote was taken at the General Assembly concerning the non-implementation of this fund; buildings in which a DTG has been carried out and that this one does not show any need for work within 10 years.

[80] BOUYEURE Jean-Robert, "Le fonds de travaux", *Informations Rapide de la Copropriété*, n°628, May 2017

This is a way to partially compensate for the negligence of people who do not care about the future. This helps to alleviate some of the negligence of people who do not care about the future.

The legislator, through these different measures (dated statement, extranet, DTG, works fund, works plan), has therefore wished to alleviate one of the main causes of financial difficulties: the lack of knowledge of future expenses on the part of the co-owners. Only the short-sightedness of the purchasers81 and the expenses linked to an external element (storm or other) can be only partially taken into account. If buyers are not allowed to use the extranet, due to a lack of confidentiality, and even if this is part of the real estate agent's duty to advise, a simple questionnaire could be envisaged to be filled out by the buyer to ensure that he is aware of the functioning of a co-ownership and its future obligations. Another questionnaire of this type could be implemented to ensure that the information provided to the co-owners is truly understood. In addition, to reinforce its objective, the legislator wished to mitigate another cause of financial difficulties due to co-owners: the absence of participation in general meetings.

### I.1.2.2 A sudden desire to combat low attendance at general meetings

Certain resolutions at general meetings must reach a quorum82 and a certain majority83 , otherwise the resolution cannot be accepted, which can indirectly lead to financial difficulties. Indeed, according to Dominique BRAYE, author of the report "Preventing and curing difficulties in co-ownerships

*"Absenteeism at a general meeting has the disadvantage of facilitating blockages and paralysing the taking of decisions which are important for the operation and maintenance of the property* "[84]. Aware that absence from a general meeting may result from the impossibility for the co-owner to be free on the day of the meeting, the legislator has recently authorized the possibility of attending by dematerialized means, of being represented or of voting by means of a pre-filled form. These devices are effective in increasing the number of voters, but can cause problems that can indirectly lead to financial difficulties.

---

[81] Misunderstanding or not reading these documents.

[82] Minimum number of voters, present or represented, required to validate a decision.

[83] In recent years, there has been a relaxation of the majorities required.

[84] ROUX Jean-Marc, Prévenir et redresser les copropriétés en difficulté, Edilaix, Point de droit, p.250, 2016

Resulting from the ELAN law, article 17-1 A of the 1965 law now authorizes co-owners to *"participate in the general assembly by physical presence, by videoconference or by any other means of electronic communication allowing their identification"*. The question of geographical distance is therefore no longer an excuse for not participating in these meetings. However, there are limits to this system. According to Laurence Guéguan-Gelinet, *"there is a risk that the debates will be partially inaudible, either because of a large number of speakers sometimes speaking at the same time, or because of more or less long interruptions due to a poor connection "*[85] . 85 However, it is necessary to have a perfectly clear debate during the GA to avoid any misunderstanding and any unwanted vote, which would be conducive to subsequent unpaid bills. It also fears *"the unsuitability of this type of transmission for small co-ownerships administered by voluntary trustees who could not deploy the necessary equipment "*[86].
Voting by correspondence, via a form87 , also has its limits. Indeed, article 17-1 A of the law of 1965 specifies that forms which do not give any precise meaning to the vote or which express an abstention are considered as unfavourable. "The *legislator's objective of protecting the co-owner is well understood, since if he is considered as favourable, or even abstaining, on the proposed resolution, he is deprived of any right of appeal "*[88]. On the principle, this is a good initiative, however, for resolutions which change substantially during the GA, the votes of the co-owners who have used the form cannot be taken into account. There is another major disadvantage. If all the owners vote by correspondence (except for one owner, since a chairperson of the meeting who is a member of the co-ownership is necessarily required).

---

[85] GUEGAN-GELINET Laurence, "Loi ELAN et Copropriété", *La revue des Loyers,* n°933, January 2019
[86] GUEGAN-GELINET Laurence , cited above
[87] Authorized by paragraph 2 of section 17-2 A of the 1965 Act
[88] LEBATTEUX Agnès, "L'amélioration des " modalités de gestion de la copropriété " par la loi Elan ",
*Loyers et Copropriété,* n°2, February 2019

There is no longer any debate, which can harm the good management of the building and may indirectly cause financial difficulties. In these situations, the surveyor acting in the capacity of syndic will see his duty and his responsibilities amplified. Furthermore, the ARC expresses the fear that "*the co-owners adopting this approach will turn to low cost proposals contrary to the interests of the syndicate* "[89]. The exchange with the trade union council makes it possible to avoid that the co-owners fall into the easy way out by taking the cheapest. This new voting system created to fight against absenteeism at the general meeting can certainly be a good alternative to the impossibility of attending a general meeting, but it must remain exceptional so as not to imply voting decisions contrary to the economic interest of the syndicate.

The last provision aiming at palliating the impossibility of presence at this annual meeting is provided for in article 22 of the law of July 10, 1965 and authorizes the delegation of the right to vote. Recently softened by the ELAN law90 , this is a practice which remains nevertheless dangerous. Indeed, in case of receipt of blank powers by the syndic, even if article 8 of the decree of 27 June 2019 has provided for the obligation to hand over these powers at the beginning of the meeting to the president of the trade-union council, "*it is possible to be tempted to distribute the powers as one pleases, in order to manipulate the votes in one's favor* "[91]. This is why it is advisable to attach to this blank power of attorney the votes to be cast on each question on the agenda. This makes it possible to reduce the risks of disputes that could lead to subsequent unpaid votes.

The recent provisions allowing to fight against the absence of participation in the general assemblies are therefore welcome, but must remain an exceptional alternative in order not to risk the appearance of conflicts of interest, likely to generate subsequent unpaid bills. However, even if efforts have been made to allow the co-owners to participate in these meetings, some do not participate or only go to oppose heavy works despite the state of the building. Furthermore, in front of a In such a situation, the solutions to be considered to fight against these indelicate co-owners are less.

---

89 "Les difficultés que pose le vote par correspondance", *La Revue de l'ARC et de l'UNARC,* pp. 24-27, n°123, 1st quarter 2019

90 A co-owner may receive more than three delegations if the total number of votes he has does not exceed 10% of the votes as opposed to 5% before

91 GUEGAN Laurence, "Différents motifs d'annulation de l'assemblée générale de copropriété", *La Revue des Loyers*, n°938, June 2013

### I.1.2.3 The fight against dishonest co-owners: ineffective measures

When a co-owner is not invested in his property, he very rarely undertakes works which can have an impact on the whole of the co-ownership making it in financial difficulty in the long run. These non-invested co-owners are regularly "sleep merchants", i.e. people who acquire apartments that they will then rent out. Although they are most often found in condominiums already in financial difficulty92 , it is not impossible to observe their presence in condominiums that are currently financially healthy. In order to prevent the financial deterioration of the co-ownership due to the latter, the legislator has provided for measures aimed at identifying and sanctioning them. However, these measures are not very effective and are often circumvented.

In view of their fraudulent activity, the presence of slum landlords can generate difficulties for the durability of the co-ownership since they try to make the apartment a potential source of income often in spite of the state of the building. Moreover, once the building is degraded, they very seldom engage rehabilitation works and thus very often vote against. Some bodies have been created to identify these situations, such as the Pôle Départemental de Lutte contre l'Habitat Indigne (PDLHI), but this remains a difficult task. "*There is not really a profile corresponding to these investors* "[93]. *93* If they cannot be identified at the time of acquisition, it is possible to report them once awareness has been raised. This reporting will allow the State to simply seize their lots without expropriation compensation. In addition, they will no longer be able to purchase housing for 10 years for purposes other than its personal occupation.95 However, the implementation of the tool requires knowledge of the facts.

---

[92] What is holding back its recovery

93BRESSON Réné, TULIER-POLGE Florence, MENJUCQ Michel, " Les copropriétés en difficulté-Table ronde ", *Revue des procédures collectives* n°1, 2017

[94] The ELAN law has modified article 18-1-1 of the 1965 law, from now on, professional trustees, property managers and real estate agents must report to the public prosecutor any facts likely to constitute an offence provided for in articles 225-14 of the Penal Code, L. 1337-4 of the Public Health Code and L. 123-3, L. 511-6 and L. 521-4 of the Construction and Housing Code.

It is therefore understandable that the response may take a long time to be put in place. In addition, in order to reinforce the measures to identify the latter, the communities have the possibility of imposing on the co-owners concerned, the holding of a permit to rent96 or a permit to divide97. Introduced by the Alur law, the rental permit allows a municipality to impose administrative procedures on a co-owner who plans to rent out a dwelling. The purpose of the permit to divide is to inform local authorities of the availability of a single dwelling for the benefit of several households. Nevertheless, "*the creation of additional dwellings without prior work, the method most commonly used by slum landlords, is not subject to any authorization* "[98] . In addition, according to the legal department of UNIS, a real estate union, "*the local authorities do not have the necessary means to carry out physical checks*". In addition, the fact that there is a fee to be paid may scare off honest investors. It is therefore wise to leave the choice of adopting such an approach to the communities. This makes it possible to concentrate on the most fragile territories. Since the rue d'Aubagne accident in Marseilles,99 several communities have adopted this system, but the number of communities that have adopted it remains marginal.100 Moreover, despite these systems, fewer than 100 convictions per year are handed down to these unethical co-owners, whereas the number of substandard housing units is estimated at more than 420,000.101 This is why, in order to increase the efficiency of these measures aiming at identifying the indelicate co-owners who are the bearers of a potential financial difficulty, it is interesting to give more weight to the tenant occupants. They can currently lodge a complaint against the authors of the fraudulent renting, nevertheless, in front of the concern that they lose their status of tenant, it is necessary to ensure them a maintenance in the premises or relocation so that they can denounce their owners.

---

[95] Circular of February 8, 2019 on the reinforcement and coordination of the fight against substandard housing

[96] Article L 634-1 CCH

[97] Article L 111-6-1-1 CCH

[98] EPF Ile de France, http://www.luttecontrelesmarchandsdesommeil.fr/une-priorite/, accessed 03/2020 [99] A balcony collapsed on November 5, 2018, killing 8 and injuring 2; https://www.leparisien.fr/faits- miscellaneous/building-collapses-a-marseille-a-balcony-collapses-pending-white-market-10-11-2018- 7939590.php

[100] Approximately 30 communities; https://www.dossierfamilial.com/actualites/immobilier-quelles-villes-ont- insta-leave-to-rent-400045, 2019, accessed 12/02/2020

101 ROUX Jean-Marc, "Le syndic et les "marchands de sommeil", *Informations Rapides de la Copropriété, CCED n°15,* 2018

In parallel to all these elements, if a co-owner is not up to date with his charges in the sense of 2° II of the law of 1965, it will be impossible for him to acquire another lot in this same co-ownership. It would be interesting to extend this system so that, when a person acquires a lot, the notary knows whether or not this person is already indebted to another building, in which case it will be impossible to conclude the sale.
There is therefore a will on the part of the legislator to counter these sleep merchants, but the tools at their disposal are lesser and only allow to partially solve this potential cause of financial difficulties of co-ownerships.

The multiplication of measures against the indirect causes of financial distress are certainly late in coming, but they allow the number of these failing co-ownerships to be reduced. The legislator is aware that these measures do not allow to totally stop a financial difficulty, and has therefore foreseen measures allowing to identify and to accompany the co-owners at risk so that they do not become part of a process of de-qualification.

### I.2 Alert systems, a use that deserves to be promoted

"*Local authorities are becoming increasingly aware that if they do not take preventive action, part of their private housing stock is likely to be in great difficulty and would be very costly to repair* "[102]. 102 To take preventive action, they must identify at-risk condominiums and provide them with support. Although identification is not always easy, these preventive actions are nonetheless effective and make it possible to reduce the number of co-ownerships in financial difficulty.

The mechanism is therefore based first of all on the necessary identification of at-risk condominiums, which is carried out through the condominium registration register or through indicators that can be provided by the Condominium Watch and Observation system (VOC).

---

[102] GELLY Ozone, "Un POPAC pour consolider les acquis et préparer le retour à un fonctionnement normal de la copropriété", *Les Cahiers de l'Anah,* n°148, April 2016

The condominium registry103 created in 2014, lists various information on condominiums104 "*allowing for the adaptation, implementation and evaluation of public policies on condominiums or the reduction of substandard housing* "[105]. One of the information to be taken into consideration in order to adapt a good prevention policy is that of unpaid bills106 . However, some condominiums are not yet registered. According to

At the ANAH, only 54% of co-ownerships would be registered. For the ARC, the remaining 66% would be

"The most common types of *co-ownerships are those with less than ten lots, those without a trustee, or those where management is chaotic or even absent*. A thorough search of the latter should therefore be carried out to ensure that this register fulfils its purpose. A first initiative consisting of depriving a defaulting condominium of a subsidy was set up in 2014108 in the hope of reducing the number of condominiums not registered in this register, but a field study should be carried out to be truly effective109. Moreover, if one looks only at the rates of unpaid debts, this may hide a more critical reality. This is why, in the absence of precise and reliable data on the entire territory, it is possible to compensate for this lack of knowledge of the state of the condominium stock in specific sectors. Indeed, the ANAH in 2012, had left the possibility to the communities to equip themselves with the VOC device. Based on statistical and field data, the VOC helps communities to identify early signs of fragility based on indicators related in particular to the functioning, management and income of the inhabitants of a condominium. After having identified, with varying degrees of difficulty, co-ownerships showing signs of fragility, the community can implement the POPAC program.

---

[103] Article L 711-1 CCH

[104] Condominiums intended exclusively for commercial or professional use escape the obligation since Article L.711-1 of the CCH only obliges "syndicates of co-owners... who administer buildings intended partly or wholly for residential use"

[105] https://www.anil.org/immatriculation-coproprietes/, accessed on 12/12/2019

[106] For example, in Saint Malo, the register of co-ownerships identifies 16% of co-ownerships (163) with more than 25% of unpaid debts.

[107] ARC.fr, L'immatriculation de l'ensemble des copropriétés : on est encore loin du compte, 03/05/2019

[108] Article L 711-6 CCH,

[109] Article L711-5 of the CCH requires the notary who encounters an unregistered co-ownership to register it, but this provision only partially resolves the lack of registration.

"*This system, which requires going into the field, completes the knowledge provided by the indicators statistics of the VOC* "[110] and aims at "*intervening with the co-owners as soon as the first weaknesses appear* "[111]. This physical accompaniment allows, in addition to giving advice, to make the co-owners as responsible as possible, thus avoiding difficulties as much as possible. "*The actions may consist of reviewing the legal organization of a co-ownership, renegotiating contracts, and making the co-owners aware of their rights and duties. By intervening before the difficulties become too serious, the POPAC can stop the deterioration and avoid communities having to resort to much more costly measures* "[112]. Moreover, the generalization of these measures was adopted by the Anah Board of Directors on November 25, 2015. This decision was made "*in light of the positive results obtained from communities that have experimented with these devices since 2012*"[113]. However, despite the positive feedback from communities, as of September 1, 2015, only 8 Voc and 24 Popac were committed. 28 Voc and 37 Popac were under development. According to ANAH, the difficulties lie in the possible reluctance of syndics or certain owners to rely on the proposed scheme. This low number can also be explained by the rather restrictive eligibility conditions since the ANAH does not allow co-ownerships with a rate of more than 75% of secondary residence to benefit from the program.

In spite of their late arrival, there are numerous mechanisms for preventing condominiums in financial difficulty and they are quite effective if the actors of the condominium are involved. If it is not possible to prevent the financial difficulty, it is necessary to look for solutions to allow the co-ownership to return to a normal situation.

---

[110] GELLY Ozone, "Un POPAC pour consolider les acquis et préparer le retour à un fonctionnement normal de la copropriété", *Les Cahiers de l'Anah,* n°148, April 2016

[111] Agence nationale de l'habitat and Ministry of Territorial Cohesion and Relations with Local Authorities, "Plan Initiative copropriétés Des dispositifs pour accompagner les interventions locales", 2019 [112] DORMOIS Rémi, "Le POPAC peut éviter de recourir à des opérations lourdes" *Les Cahiers de l'Anah,* n°148, avril 2016, p.13 ; *Local authorities also use the POPAC to support the recovery of condominiums that have benefited from measures such as the Opah Copropriété, the Safeguard Plan or urban renewal. It also makes it possible to verify that the initial provisions of these projects have been achieved.*

[113] DORMOIS Rémi mentioned above

## II The recovery of condominiums in financial difficulty, a process to be undertaken quickly

The surveyor intervening in the capacity of trustee must examine the real needs of the co-ownership before initiating measures to rectify the situation. In certain situations, the correction of a defective element of the co-ownership can be sufficient to durably rectify the situation. In other cases, the co-ownership is failing because it is facing a serious lack of cash flow. Finally, in more complex cases, the co-ownership is in difficulty because it requires the accumulation of the two previous needs to recover. Thus, three levels of difficulty emerge and correspond to the levels of difficulty evoked by legal practitioners: fragile co-ownership (II.1), co-ownership in difficulty (II.2) and degraded co-ownership (II.3). The law of July 10, 1965 does not retain these three levels, but implies them through the possibility of resorting to three types of support: the ad hoc representative, the provisional administrator and the reinforced provisional administrator. It is indeed necessary to identify the level of financial difficulty as well as the related needs, otherwise the recovery to a normal situation will be strongly compromised. Following this examination, it will be advisable to rely on the law governing co-ownership as well as on the State because they bring relatively effective devices to answer these needs, but the more the state deteriorates and the more the return to a normal situation will be long, even impossible.

### II.1 Fragile condominiums: a need to clean up governance

A fragile co-ownership is considered to be "a *co-ownership showing signs of fragility, at risk of entering into a spiral of de-qualification which could eventually lead to difficulties* "[114]. Although there are other indicators of fragility, also based on the need to clean up governance115 , unpaid bills remain the indicator most often used by legal practitioners, since this can put the co-ownership in danger if the situation drags on.

---

[114] DEVRY Pascal, " Le point sur : The new ANAH aid for fragile condominiums", *cahier des copropriétés en difficulté,* n°11, p.2, 2017

[115] Presence of lax syndic, Age of the building, Rate of secondary residence, energy classification of the building, percentage of presence at the general meetings...

The ANAH, which is the main organization that can deliver subsidies towards certain co-ownerships considers that a co-ownership presents signs of fragility as soon as it displays an amount of unpaid charges between 8 and 15% of the provisional budget and between 8 and 25% of unpaid charges for co-ownerships of less than 200 lots. It is judicious to consider unpaid bills as an indicator of fragility because "*the failure of a significant part of the co-owners to pay their charges deprives the syndicate of vital resources to provide for the conservation of the building, the condition of which cannot fail to deteriorate* "[116] . *This is* particularly true since a solidarity clause between a co-owner who is in debt and the other co-owners is deemed unwritten by case law117. It is nevertheless accepted that the debts of a co-owner who is in default may be paid by the other co-owners if it is decided by the general assembly (this may, in fact, avoid, for example, water or other cuts)[118] . However, most of the time, the other co-owners are unable or unwilling to pay this debt119 especially when the amount of unpaid bills is significant or when this amount can only be divided among a few co-owners. All the more so when they do not necessarily have a precise timetable for the return of the sums advanced. Also, the presence of an unpaid amount, if it lasts, is conducive to a spiral of degradation which it is necessary to put an end to. To ensure this rapidity, coupled with the different methods that can help to regularize the situation (II.1.1), it is possible to compensate for the deficiency of certain trustees120 or co-owners by appointing a person to assist them (II.1.2).

---

[116] GUILHEM Gil, "Copropriétés en difficulté - Garanties de recouvrement et syndicats de copropriétaires en difficulté", *JurisClasseur Copropriété*, Fasc.84, February 2018, spec. n°1

[117] DRAY Yoann, La défaillance d'un copropriétaire dans le paiement des charges, *legavox.fr*, 2012 *: Article 10 of the law of July 10, 1965, which concerns the payment of condominium fees, is of public order in all its provisions. Thus, a clause exempting certain co-owners from the payment of charges or imposing on a co-owner the charge of an item of equipment or a service which would not be of any use to him is deemed unwritten (Law of 10-7-1965 art. 43, al. 1); see also CA Paris March 11, 1982 + CA Versailles 23.03.1989*

[118] CA Paris, 23rd ch. March 4, 1986, commented by Inf. rap. copr. sept. 1986, p. 156

[119] TOMASIN Daniel, "Further reflections on co-ownerships in difficulty", *Droit et Ville,* n°73, pp 11-18, 2012

[120] reference is made to the notion of the trustee's failure to act in articles 18 V of the law of July 10, 1965 and 49 of the decree of March 17, 1967 and through various jurisprudences.

### II.1.1 The development of unpaid bills, a situation to be regularized

The surveyor acting in the capacity of syndic must be aware that the collection of charges, which is an integral part of his mission121 , is an often long and sometimes doomed to failure procedure. To reduce this failure, in addition to the collection of traditional charges (II.1.1.1), he must keep in mind that the division of the syndicate can help to stop these unpaid charges (II.1.1.2).

#### II.1.1.1 The recovery of charges, a particularly long procedure in the case of indelicate co-owners

Although the syndic has a specific procedure provided for by the law of co-ownership for the recovery of the charges which takes place in summary proceedings, it turns out that to date, the injunction to pay is the most used recovery device. Moreover, after having waited for a long time for the writ of execution justifying the forced execution, the recovery is not always complete.

When a debtor has not paid because of a temporary lack of cash flow, or because he is dragging his feet, the trustee can set up a payment schedule. This will make it possible to spread out the amount due over time, allowing the debt to be recovered without having to resort to heavier collection procedures. In addition, before any legal action is taken to recover an unpaid debt, the debtor must attempt to collect the debt out of court.122 This action can be taken "as soon as the debtor is aware that the debt has been paid. This action can be initiated "*as soon as the first unpaid expenses are due following the vote on the provisional budget voted by the General Meeting* "[123] . After sending a formal notice to pay by registered letter with acknowledgement of receipt to the person concerned, a period of 30 days from the receipt of the letter must be respected124 before launching any legal proceedings.

---

[121] Civ. 3e, Oct. 8, 2015, pourvoi n°14-19.245, commented by ROUQUET Yves, *Dalloz actualité*: L'action en recouvrement des charges incombe au seul syndic

[122] Articles 56 and 58 of the Code of Civil Procedure require the creditor to attempt amicable collection before initiating legal proceedings

[123] VIGNERON Guy, " Charges communes-Recouvrement des charges-Procédures et garanties de recouvrement " *JurisClasseur Copropriété*, Fasc. 75-10, April 2010, speculative n°9 ; See also, Cass, civ.3 01.12.2010 n°09-72.402 for the payability of charges

[124] Article 19-2 of the law of 1965

This is an essential step since it can in many cases, In addition, it is not uncommon to find a clause in a condominium by-law which provides for an increase in charges in the event of unpaid bills and "this letter will serve as the starting point for the calculation of interest on arrears. In addition, it is not uncommon to find in a condominium by-law a clause of aggravation of charges in case of unpaid bills and "*this letter will serve as the starting point for the calculation of the late payment interest* "[125] . Although this clause is intended to anticipate the development of unpaid charges by dissuading their emergence, it remains modestly effective, as it can only be applied after a judicial finding126. If after this 30-day period the situation does not change and there is no response from the debtor, the trustee must resort to legal proceedings. The purpose of this procedure is to obtain a writ of execution for the payment of the debt. It is therefore important not to delay sending the letter to avoid delaying its collection even more.

After having ensured that his claim is certain (he will have to provide proof), liquid (the amount must be determined) and due (the limitation period must not have expired)[127] , an action before the judicial court may finally be possible. Obtaining the writ of execution can be done via several procedures: the procedure specific to co-ownership128 (also called the "super recovery procedure"), the injunction to pay procedure129 , or the simplified procedure130. The multiplication of these procedures invites the syndic to question himself on the best procedure to initiate in order to judiciously redress the co-ownership. The choice of the type of collection procedure to be initiated depends on the situation. In the case of small debts, the simplified procedure appears judicious. In the case of larger debts, the injunction to pay appears more appropriate. If one is faced with recurrent unpaid debts, or with a debt that needs to be recovered urgently, the procedure specific to co-ownership is the best option offered to the trustees. However, the trend shows that the injunction to pay, although not the quickest means of recovery, is the most used recourse131 .

---

[125] BERTEAUX Alexandre, Le recouvrement des charges impayées de copropriété, *Le figaro*, 2017

[126] Cass [3e] civ, 27mars 2013, n°12-13.012 : Administrer juin 2013, p.54, obs J.R BOUYEURE

[127] Article L111-2 code of civil enforcement procedures; These are relatively simple conditions to meet. These are relatively simple conditions to meet, but it is necessary to ensure that they are met. For example, the minutes of a general meeting approving the accounts are insufficient to justify the claim.CA Aix en Provence, 11th ch. 28/02/2017, n°15/21567

[128] Article 19-2 of the 1965 law governing co-ownership

[129] Article 1405 Code of Civil Procedure

[130] Article L.125-1 and R.125-8 of the Code of Civil Enforcement Procedures (formerly Article 1244-4 of the Civil Code).

[131] No specific data on this, but many legal practitioners agree.

The explanation for this fact lies in the effectiveness of the of the procedure132 and its ease of implementation unlike the other two. Indeed, the injunction to pay only requires to provide the judge with the proof of the debt. Moreover, contrary to the simplified procedure, it is a free procedure up to 10,000€ and is not restricted by a threshold of unpaid debts.133 "*The simplified procedure aims more at relieving the courts than at really giving an additional tool to co-ownerships in the fight against unpaid debts* "[134]. As for the specific procedure for co-ownership, before it was reformed, it required the multiplication of procedures to recover all unpaid debts. The ELAN law reformed this specific procedure135 to compensate for the inefficiency of this system. There is no hindsight yet, but this reform is a major step forward in the fight against condominiums in difficulty in general, since if the trustees take ownership of this system, unpaid debts will be recovered more quickly and more efficiently.

Once the writ of execution has been obtained, the creditor can have it enforced by a bailiff in order to obtain payment of the charges by seizing the debtor's property (and sometimes selling it at auction). In case of a co-owner having serious financial difficulties, the latter will be forced to proceed to the alienation of his real estate. This requires the authorization of the GA136 which may represent an additional obstacle to the rapid recovery of the debt in the case where the GA does not wish to engage in this type of procedure. Fortunately, the debtor concerned cannot participate in the vote at this AGM137. In order to ensure that this alienation benefits the trade union, the legislator makes the syndicate of three legal guarantees.

---

[132] GUEGAN-GELINET Laurence, " Le contentieux de recouvrement des charges de copropriété après les lois ALUR et ELAN ", *Transversale immobilière,* pp.12-18, n°142, mai-juin 2019

[133] Article R125-1 Code of Civil Enforcement Procedures: above 4000€, the use of the simplified procedure is no longer possible

[134] Coproconseil.fr, Simplified debt collection Macron law, March 2016

[135] Article 19-2 of the law of 1965 originally made it possible, in the event of non-payment of a single call for charges, to make all the other quarters of charges for the current accounting period payable. Today,
*"This "situation" has been extended to include unpaid amounts from previously approved fiscal years. This makes it possible to initiate only one procedure against a co-owner in bad faith to recover both previous charges and future provisional calls.* »[135]. ARC-COPRO, La déchéance du terme version loi ELAN : " une super procédure " de recouvrement des charges, May 2019

[136] Decree of March 17, 1967, art. 55, para. 2

[137] Article 19-2 law of 1965

The mortgage,138 the special real estate lien139 and the movable lien.140 These guarantees are intended to enable the syndicate to facilitate the recovery of the debt since the syndicate, in the event of the sale of the real estate of the defaulting co-owner, will be given preference over other creditors for the current year and the last two years due.141 However, "the procedure leads to a public auction of the debtor's property, but the price is often derisory in relation to the amount of the debt. However, "*the procedure leads to a sale by public auction of the debtor's property, but the bidding is often derisory in relation to the amount of the debt* "[142]. 142 Moreover, the joint and several clause between a buyer and a seller is deemed unwritten.143 This fact shows us the interest of this type of procedure. This fact shows us the interest for a trustee to act as quickly as possible in order to avoid the amount of the debts being too great.

In order to compensate for "*the slowness of the justice system, which develops the feeling of impunity of the defaulting co-owners, and discourages the others who have to advance the money in their place* "[144] , the ELAN law has come to reinforce the lack of efficiency of the specific procedure for co-ownership. In addition, in order to allow for a rapid recovery, it is possible to envisage the use of a division of the syndicate.

#### II.1.1.2 The division of the union, a potentially effective alternative to counter the situation of the development of unpaid bills

In the presence of large condominiums with several independent buildings, it is not uncommon, as we have seen, to face a management problem that causes financial difficulties. Moreover, "*solutions have been piled up, without solving the real problems* "[145]. This is why, in the absence of preventive action (I.1.1) and despite.

---

[138] Article 19, paragraph 1, of the law of July 10, 1965. This applies to claims of any kind that the syndicate holds against the said debtor co-owners, whether they are provisional or final payments
[139] Article 19-1 of the law of July 10, 1965
[140] Article 2332 Civil Code
[141] Article 2374 Civil Code
[142] M. Lang Pierre, Member of Parliament for Moselle during a question to the Union pour un Mouvement populaire Group
[143] Cass. 3e civ., July 1, 1980 : Bull. civ. 1980, III, n° 127
[144] M. Lang Pierre, op. cit.
[145] TOMASIN Daniel, "Further reflections on co-ownerships in difficulty", *Droit et Ville*, n°73, pp 11-18, 2012

*In addition to the* disadvantages seen previously146 , "*Law no. 2014-366 of March 24, 2014, known as the Alur Law, sought to make the division of the syndicate a general device for the recovery of co-ownerships in difficulty by opening up the possibilities of constituting a secondary syndicate or of dividing it, including in volumes, to the judge who pronounces it* "[147]. This recourse may make it possible, in addition to avoiding a recurrence of unpaid debts, to divide the liabilities of a co-ownership in order to accelerate the recovery process.

As mentioned above, the purpose of dividing the syndicate is to allow the co-owners to feel more involved in the work voted on at the GA148 because it will affect
"This allows them to project themselves more easily on the good functioning of the co-ownership. It is a question of palliating a management constraint implying a financial difficulty. In addition to the side linked to the re-establishment of management, in case of difficulty in collecting the charges, the division of the syndicate can prove interesting for the fate of the debts and receivables. Indeed, the provisions of article 28 of the 1965 law specify that debts are attached to the lots149 . This division will make it possible to accelerate the recovery process. Indeed, "*after the distribution of the debts, the new entity will benefit from its own recovery plan* "[150] . This new and more coherent scale of recovery plans aims at making it easier for the co-owners to accept the measures
"In addition, "if the co-ownership as a whole does not prove to be recoverable, the separation of debts and liabilities between the secondary syndicates will simplify the implementation of the deficiency procedure. Moreover, "*if the co-ownership as a whole does not prove to be recoverable, the separation of debts and claims between the secondary syndicates will simplify the implementation of the deficiency procedure* "[151] . In addition, to facilitate recovery due to unpaid debts, the division of the syndicate may allow for a reduction in charges. Indeed, certain parts can be transferred to the commune or to a developer who will ensure their maintenance.

---

[146] Size of the amount to be committed, need for an agreement at a general assembly, need to have the required conditions, length of the procedure

[147] TOMASIN Daniel, "La Copropriété", *DALLOZ Action*, chapter 325.181 Syndicat en difficulté, 2018/2019

[148] Thus reducing blockages due to votes and unpaid charges due to disputes.

[149] The splitting of a co-ownership does not entail the liquidation of claims because this would imply the disappearance of the syndicate; answer of the Ministry of Housing in 2004 to a question asked by Mr. JEANJEAN Christian, deputy of the Hérault

[150] Private Housing Policy Forum, Association syndicale professionnelle d'administrateurs judiciaires, hearing of m° TULIER-POLGE of ASPAJ, Sept. 30, 2015

[151] The above-mentioned forum

It will nevertheless be necessary to vigilant to possible conflicts of interest.152 For example, after acquisition, the commune may wish to leave a former common area of the condominium open to access. This new destination may upset certain co-owners, which may hinder a lasting recovery. According to Me LEBATTEUX153 , the division may also allow for the realization of real estate transactions. It is indeed interesting to take advantage of this division to mobilize the potential of the land as we have seen154.

In view of the different elements mentioned, the division of the syndicate is a measure which can allow the anticipation of financial complications, but can also, even if it is not directly a recovery mechanism and is difficult to implement, accelerate the recovery process which is particularly long in the case of a co-owner in bad faith. It is therefore a mechanism that can be interesting to rectify a co-ownership in financial difficulty when its situation results from a management problem. Indeed, it could allow for the simplification of the governance and the improvement of the recovery strategy. In order for it to be effective, it is nevertheless necessary to define as precisely as possible the material, legal and financial conditions, which is not always simple and reduces the effectiveness of this mechanism. This restructuring may also appear necessary to avoid that the co-ownership falls back into a difficulty after recovery.

All of the recovery mechanisms mentioned so far, show us that through the possibility of initiating a collection procedure or the possibility of dividing the syndicate in order to hope to recover unpaid amounts more easily, that the legislator has multiplied the measures aimed at recovering fragile co-ownerships due to the presence of unpaid amounts. On the other hand, apart from the possibility to include a clause of aggravation of the charges in the co-ownership rules, no sanction is foreseen to dissuade the birth of an unpaid debt which, as a reminder, is the main indicator of the fragility of a co-ownership. However, "*in Spain, article 15-2 of the law of July 21, 1960 provides for the withdrawal of the right of voting against the defaulting co-owners.*

---

[152] TOMASIN Daniel, "La Copropriété", *DALLOZ Action*, chapter 324-325 Syndicat en difficulté, 2018/2019

[153] Lawyer at the bar of Paris,

[154] See paragraph I.2.2

*The purpose of this sanction is to make the debtor co-owners understand the importance of complying with their obligations. Although this sanction has its limits155 , it does exist. It is obvious that in France there is a lack of intermediate sanctions against defaulting co-owners "*[156]. Moreover, in the case of co-owners in real financial difficulty or in bad faith, the recovery can quickly become tedious for the syndic. The latter do not all have specific training in the financial recovery of co-ownerships. Aware that recovery requires the intervention of the syndic, the legislator wished to compensate for the possible failure of the latter by authorizing the use of an *ad hoc* representative or a provisional administrator.

### II.1.2 Assistance to the trustee: a late but recognized mechanism

As a reminder, a fragile co-ownership is characterized by a failing element of the co-ownership. This failure, which can lead to financial difficulties, may be due to the failure of the syndic. "*The jurisprudence has had the opportunity to specify the contours of this notion. This category includes, in particular, abstention from exercising legal actions in the name of the syndicate157 , negligence in the execution and payment of work158 , lack of diligence in executing a judicial decision159 , failure to execute the decisions of the general meeting160 , etc. "*[161] Failing to act preventively, it will be necessary to appoint assistance to the syndic. This is firstly the *ad hoc* representative who can nevertheless only make an observation and give possible solutions to remedy the bad situation (II.1.2.1). Secondly, it can be a provisional administrator who will be able to decide in their place in case of deficiency on the part of the co-owners (II.1.2.2). These individuals are effective in redressing the situation, but the threshold for engagement is potentially irrelevant.

---

[155] It may seem unfair to penalize a person of good faith who exceptionally encounters a financial difficulty; a co-owner who is disinterested in the system of co-ownership, and who therefore does not regularize his expenses, will not care if he withdraws his vote

[156] INFANTI Maeva, *Étude des difficultés juridiques inhérentes à la présence d'une copropriété dans ses relations avec les tiers,* Mémoire de Master Identification, Aménagement et Gestion du Foncier, ESGT, p.82, 2014

[157] Cass. 3e civ., 19 Apr. 1974 : JCP N 1975, prat. 6013.

[158] CA Paris, June 9, 1999: JurisData n° 1999-102202; RDI 1999, p. 461, obs. Capoulade; Dossier Csabn° 78, Nov. 1999, n° 133; IRC March 2000. 12, note Capoulade.

[159] CA Paris, Jan. 8, 2003: AJDI 2003, p. 425.

[160] Cass. 3e civ., May 29, 2002, n° 00-21.739: JurisData n° 2002-014501; Administrer août/sept. 2002, p. 61.

[161] IVARS Camille, "L'impact de l'ordonnance du 30 octobre 2019 sur les règles procédurales en droit de la copropriété", *Loyers et Copropriété,* n°2, february 2020, dossier 11, spec. n°19

### II.1.2.1 The ad hoc representative: an assistance not to be neglected

The role of the *ad hoc* trustee is to draw up an inventory of the situation and find solutions aimed at restoring financial equilibrium162 , but in order to be able to call upon him it is necessary to reach a threshold of unpaid debts. However, in addition to the late recourse threshold, the ad hoc representative has only limited power within the framework of a mission that will only be one-off and specific, which hampers the effectiveness of this mechanism.

Provided for since 2009 by the MOLLE law, the procedure to resort to an *ad* hoc representative has been modified by the ALUR law. The latter has decreased the threshold of unpaid necessary from

This procedure can only be initiated by a "meeting" of co-owners. A co-ownership may now request the appointment of an ad hoc representative as soon as it is faced with a rate of unpaid debts higher than twenty-five percent for co-ownerships comprising less than two hundred lots and fifteen percent for other co-ownerships163. The choice of a threshold was justified in order to avoid abusive recourse. In addition, *"in order to avoid triggering an alert procedure for simple late payments of no consequence, the texts specify that sums which became due in the month preceding the closing date of the financial year are not to be considered as unpaid "*[164]. However, the threshold undertaken may in certain cases prove harmful to the co-ownership. In fact, in a small condominium, recourse to such assistance should be initiated even before reaching fifteen or twenty-five percent of unpaid bills. In spite of the assumption that in the near future the situation will deteriorate without a wise advice, a certain number of co-ownerships cannot resort to this type of assistance. This concerns for example the anticipation of heavy works. In France, contrary to Germany, a co-owner who fails to meet these obligations can be evicted by the other co-owners who can demand the sale of his lot.165

---

[162] Article 29-1 B law of 1965

[163] Article 29-1 A law of 1965

[164] GUILHEM Gil, "Copropriétés en difficulté - Garanties de recouvrement et syndicats de copropriétaires en difficulté", *JurisClasseur Copropriété*, Fasc.84, February 2018, spec. n°44

[165] Articles 18 and 19 of the law of 15 March 1951 on the ownership of apartments (Gesetz über das Wohnungseigentum und das Dauerwohnrecht (Wohnungseigentumsgesetz)).

*"This sanction proves to be a real means of pressure for the co-owners defaulters, because disinterested or negligent persons will quickly be removed from a co-ownership which wishes to be "healthy* "[166]. Without going as far as the sale, it would be interesting to extend the above-mentioned threshold to the defaulting co-owners so that they can benefit from a personalized accompaniment167 , thus avoiding that the whole co-ownership is affected.

In order to facilitate access to this device, the legislator, with the ALUR law, has multiplied the number of actors who can refer to the court for the appointment of this agent. This wider referral allows to put more chances on one's side in order to avoid that a co-ownership does not miss this recovery aid. This referral can be made by the syndic who will inform the syndicate council, by the co-owners representing together at least 15% of the votes of the syndicate. This is a fairly high threshold, because *"it implies a quasi collective action*. The court could also be seized by a creditor169 and, since the ALUR law of March 24, 2014, by the representative of the State in the department or the public prosecutor, by the mayor of the municipality of the location of the building or finally by the president of the deliberative body of the public establishment of inter-communal cooperation (EPCI) with jurisdiction over housing, in the location of the building.

Once seized, the mandataire *ad hoc* must carry out a mission similar to that of the mandataire *ad hoc* appointed to assist in the reorganization of companies in difficulty.170 In fact, in the event of a dispute, this mandataire mediates and negotiates between the parties. He must draw up recommendations to restore financial order and submit a report on his observations and recommendations to the court. He has the power to ask the judge

---

[166] INFANTI Maeva, *Étude des difficultés juridiques inhérentes à la présence d'une copropriété dans ses relations avec les tiers,* Mémoire de Master Identification, Aménagement et Gestion du Foncier, ESGT, p.82, 2014

[167] Giving them advice to reduce their expenses, to renegotiate contracts...

[168] LEBATTEUX Agnès, "Les dispositions de l'ordonnance n° 2019-1101 du 30 octobre 2019 relatives aux modifications de la structure de la copropriété (syndicat secondaire, surélévation, scission) et aux copropriétés en difficulté", *Loyers et Copropriété*, n°2, février 2020

[169] When the invoices for water or energy subscription and supply, or the invoices for works, voted by the general assembly and executed remain unpaid for six months and the creditor has sent the managing agent a summons to pay which has remained unsuccessful. However, creditors of unpaid invoices which do not meet the conditions of article 29-1 A paragraph 3, of the law (for example, insurance premiums, indemnities due to third parties or co-owners, works not voted by the general assembly) are excluded.

[170] Article L611-1 to L611-16 of the Commercial Code

to terminate a contract or to force its execution in the interest of the syndicate, which the syndic cannot do unilaterally. Moreover, this type of accompaniment "*can be useful to promote the awareness of the co-owners when the syndic is not listened to. It is a powerful tool to know the situation of a co-ownership when the syndic does not cooperate or is in default*".[171] Nevertheless, it is regrettable that fragile co-ownerships have not benefited from the mechanism allowing them to fix their debts. The *ad hoc* trustee may therefore find it difficult to estimate the real amount of the debts.

Although useful in the texts, this device has nevertheless been little used. Only 7 requests were made in 2013172 compared to 15 in 2014173. As the ANIL points out,
*"It is a text which was a priori very interesting, but which has hardly been applied since 2009 because it had no coercive character. Moreover, many trustees did not wish to use it because they feared that the administrator's mission would consist in a pure and simple criticism of their management, relayed by the co-owners "*[174]. The use of an *ad hoc* trustee is therefore faced with a "*reluctance on the part of the various actors to implement it. It is not desired by the syndic, who fears that the report will point out his failings, nor by the commune, which fears having to assist the co-ownership "*[175.]

In view of the various disadvantages mentioned and the lack of use in practice, one may wonder whether the existence of such an accompaniment should not be reconsidered. In the absence of a mandatary ad hoc and in order to complete the powers of the latter, it is possible to resort to a provisional administrator who is more solicited than the mandatary *ad hoc* in the case of a co-ownership in financial difficulty.

---

[171Forum] des politiques de l'habitat privé, "Atelier copropriétés - Le redressement judiciaire", 2015

[172] Report by Claude DILAIN, Senator, 2013

[173] Directorate of Civil Affairs and the Seal

[174] ANIL, Co-ownerships in difficulty / Ad hoc representative and provisional administrator, 2015

[175] LEBATTEUX Agnès, "Les dispositions de l'ordonnance n° 2019-1101 du 30 octobre 2019 relatives aux modifications de la structure de la copropriété (syndicat secondaire, surélévation, scission) et aux copropriétés en difficulté", *Loyers et Copropriété*, n°2, février 2020

### II.1.2.2 The provisional administrator: definite advantages to intervene more quickly

The provisional administrator is invested with a general mission and has more extensive powers than the ad hoc representative: he has the vocation to substitute himself for the syndic and the syndicate. However, the provisional administrator will certainly respond to the needs of a fragile co-ownership in order to clean up its governance, but recourse to this type of support is not always possible in the absence of proven difficulties.

The request for the appointment of a provisional administrator, although it can be made by a multiplicity of actors, requires to meet the conditions defined in article 29-1 of the 1965 law. It is necessary that " *the financial balance of the syndicate of co-owners is seriously compromised or that the syndicate is unable to provide for the conservation of the building* ". The criterion for the opening of a provisional administration procedure is therefore no longer based on a threshold of unpaid amounts in relation to the voted budget, but results from a general situation. Apart from the absence of a trustee, it is doubtful that the provisional administrator can be seized in the case of a fragile co-ownership. Indeed, unless it can be proved that the situation will become more complicated in the near future, it is necessary to wait for a complication to resort to this type of support. "*In many cases, the appointment of a provisional administrator comes too late to be really effective176* ".

However, if it is possible to seize it, it would *allow "access to a mechanism specific to co-ownerships in difficulty, making it possible to confer on an administrator appointed by the judge extensive powers in matters of management and decision-making, in place of the general assembly* "[177]. Unlike the ad hoc agent and the trustee, he has all the powers of the trustee (whose mandate ceases by right without indemnity) and all or part of the powers of the general meeting of co-owners178 . This power will enable him to compensate for the possible failure of the co-owners to vote.

---

[176] Pierre CAPOULADRE, "Prévenir, repérer et traiter les syndicats en pré-difficulté", *AJDI*, March 2007, p.193

[177] Braye Dominique, Prévenir et guérir les difficultés des copropriétés. Une priorité des politiques de l'habitat, 2012, p.

[178] Art 29-1 of the law of 1965

Indeed, "*in the face of assemblies general blockage, placing the building under provisional administration may be necessary to undertake work essential to the conservation of the building and to deal with unpaid debts* "[179]. In addition, if certain co-owners refuse to proceed with the sale of assets of the common portions, which is the only possible solution for financial recovery, the provisional administrator may unilaterally substitute himself for them180. On the other hand,
*"The administrator will have to be trustworthy since the co-owners will feel dispossessed of their management and their money* "[181].

The other notable advantage of using such a mechanism to redress a co-ownership in financial difficulty is that the provisional administrator can liquidate the debts of a syndicate more easily. According to the provisions of article 29-3 of the law of 1965, the decision to appoint a provisional administrator suspends the payability of debts, for 12 months (which can be extended up to 30 months). This makes it possible to freeze the situation while a favorable outcome is found. Moreover, in the same idea, article 29-4 specifies that within two months of his appointment, the provisional administrator shall proceed with publicity measures to allow the creditors to produce the elements necessary to evaluate the amount of their claims. "*Thus this wording copies the rules of collective proceedings which it applies to co-ownerships* "[182] . 182 Here the creditors will have three months to declare their claims, failing which the forgotten claims can no longer be invoked183. This fixing of the amount of the claims will facilitate the task of the provisional administrator. In addition, in the absence of assets of the syndicate of co-owners that can be transferred under the conditions defined in article 29-6 or if the transfers have not found a buyer, the provisional administrator may ask the judge to partially erase the debts of the syndicate for an amount equivalent to the amount of the uncollectible claims. [184]

---

179 ARC-Copro.fr, What role can communities play in putting condominiums in their territory under provisional administration? 2017, accessed 03/03/2020

180 Cass. 3e civ., Jan. 23, 2013, n° 09-13.898 : JurisData n° 2013-000784 ; Loyers et copr. 2013, comm. 156, obs. G. Vigneron; Rev. loyers 2013 p. 150, note V. Zalewski.

181 BRAYE Dominique, supra

182 Article L. 622-24 Code du commerce.TOMASIN Daniel, "La Copropriété", *DALLOZ Action*, chapter
325.151 Troubled Union, 2018/2019

183 When a creditor fails to meet his obligation to declare his claim due to a default that is not of his own making, Act No. 2017-86 of January 27, 2017 on equality and citizenship introduced an action for relief from foreclosure in Article 29-4 of Act No. 65-557 of July 10, 1965 .

184 Art 29-7 law of 1965

These powers are a definite remedy to allow the recovery of co-ownerships with serious financial difficulties. He will be able to establish a debt settlement plan without the risk of seeing it change. This plan, including a schedule of payments to the creditors of the syndicate of co-owners, must respect a maximum duration of five years185. In the case of a fragile co-ownership, five years seems to be the right time to rectify the situation. On the other hand, for degraded co-ownerships, this period may be too short to durably rectify the situation. According to Florence Tulier-Polge, a court-appointed administrator in Evry, "*to claim to regulate debts in only five years when the rate of unpaid debts is higher than 50% is a fiction*".[186] It would be interesting to give the judge the freedom to authorize or not an exemption187 that would allow the plan to be spread out over a longer period, and to modify the 1965 law on this point accordingly.

Although these persons appear to be indispensable when it becomes necessary to thoroughly clean up the governance of the co-ownership, their missions should not be restricted to the recovery of the situation. The provisional administrator, within the framework of his mission, must be attentive to all the actors of the co-ownership. He must listen to the claims and demands of the occupants or co-owners. He must inform of the measures taken and advise the syndicate. He must therefore rectify the situation, but also make sure that the co-ownership does not fall back into a difficulty. However, the ARC deplores a recurrent lack of communication between the provisional administration and the trade union council188 . Apart from the drafting of an annual report by the provisional administrator, "*it is often difficult to obtain information from him and the documents sent are sometimes not very comprehensible. This causes a progressive demobilization of the co-owners* "[189]. According to the ARC, the administrator would also not always carry out the necessary diagnoses and would only sell the units. He should record his decisions on the register of the minutes and notify them like a minutes of a general assembly.

---

[185] Article 29-5 law of 1965

[186] ROUX Jean-Marc, Prévenir et redresser les copropriétés en difficulté, Edilaix, Point de droit, p.250, 2016

[187] Waiver to be requested only in cases of extreme necessity.

[188] ARC-copro.co.uk, What role can communities play in interim administration, 2017, accessed 03/03/2020

[189] ARC copro, supra

Certainly, the administrator constitutes a key measure to help the recovery of a co-ownership in difficulty, however, this measure requires that the co-ownership be in an already serious situation to resort to it. Reducing the conditions to benefit from a mandataire ad hoc as seen previously would allow to relieve the provisional administrators. Indeed, in certain cases, a mandatary ad hoc would be sufficient to redress the situation of the co-ownership. Moreover, once the matter is referred to him, he will be able to appoint a provisional administrator if necessary. There will therefore no longer be any need to wait for the conditions of article 29-1 of the law of 1965 to resort to a provisional administrator.

As indicated, there are numerous mechanisms to improve the governance of a fragile co-ownership. To remedy the default of a co-owner in the payment of the charges, the Elan law has facilitated the use of the collection procedure provided for in article 19-2 of the law governing co-ownership, which should allow for greater efficiency and rapidity. Moreover, although it is not always easy to implement, the division of the syndicate can accelerate the collection of charges. To compensate for the failure of the syndic or his impossibility to financially rectify the co-ownership, the appointment of an ad hoc representative or a provisional administrator is efficient, but requires to reach a certain level of deterioration. This time lag can nevertheless make their treatment complicated, which is regrettable in case of co-owners in serious financial difficulties making the co-ownership pass from a state of fragility to a state of difficulty, the latter cannot rectify the situation alone. It is then necessary to go and look for a treasury outside the co-ownership.

## II.2 Co-ownerships in difficulty: a multitude of exceptional aids

A co-ownership in financial difficulty is defined by article 29-1 of the law of 1965 as one of which "*the financial equilibrium of the syndicate of co-owners is seriously compromised or that the syndicate is in the impossibility of providing for the conservation of the building*". This situation, despite the intervention of a provisional administrator, requires to compensate for the lack of cash flow, often due to the need for financially heavy works. For this, the syndic, or the provisional administrator if necessary, has several possibilities. He must adapt to the complexity of the various public aids (II.2.1). The mobilization of land potential also appears to be a possible source of financial income, even if it is subject to restrictions (II.2.2). Finally, the recourse to financing from private persons also appears not negligible even if these financing are difficult to find (II.2.3).

### II.2.1 Public aid: important but complex

The reasons for a great financial difficulty experienced by a condominium can result from a sudden need for heavy work that is difficult to finance, but which "*is very often the only way to achieve, for the future, substantial operating savings, a lasting reduction in condominium expenses and therefore in the medium term the recovery of the syndicate* "[190]. "*Public intervention is then essential to help the lasting recovery of the condominium. This intervention can be partly financed by the Anah* "[191] , but the complexity and the strict conditions of granting reduce their effectiveness.

Faced with very heavy work from a financial point of view that puts a condominium in financial difficulty, the various financial aids of the ANAH192 (the main organization that can deliver subsidies) can help support the most modest households for renovation work on the private portions. This aid is also intended for the syndicate of co-owners for work on the common areas. A large part of this financing is devoted to energy renovation work. For a co-ownership in difficulty, the use of this type of work can have the effect of drastically reducing the energy bill and thus slowing down the deterioration of their situation. The rest of the financing aims at facilitating the recovery of co-ownerships whose state endangers the occupants. It is a significant help for these co-ownerships since the amounts granted are interesting. Despite the importance of this aid (Annex 1), it appears that it is not always used because "*the fragmentation and complexity of the aid is attributed according to different scales, rules of eligibility and procedures* "[193].

---

[190] ROUX Jean-Marc, Prévenir et redresser les copropriétés en difficulté, Edilaix, Point de droit, p.250, 2016

191https://www.anah.fr/collectivite/traiter-les-coproprietes-fragiles-et-en-difficulte/preparer-votre-

intervention/

[192] The aids in question are Ma PrimeRénov', habiter mieux sérénité, habiter mieux copropriété, Habiter sain/serein

[193] Ecologie-solidaire.gouv.fr, La ville de Metz au chevet des copropriétés privées, 2017

Moreover, "*the line between subsidizable and non-subsidizable expenses is sometimes subtle.* 194 In addition, some grants can be combined with others (Annex 2). The effectiveness of these aids is also slowed down by precise conditions for granting them195. Owner-occupiers are subject to strict resource ceilings. As for landlords, they are obliged to agree to rent their dwellings at the level of social housing. In addition, "*the ANAH no longer subsidizes work on common areas as long as it is cosmetic work, even when it is particularly degraded* "[196]. Thus, in spite of the arrival of a platform aiming to alleviate this complexity197 , "*not all co-owners or trustees are necessarily able to find their way around* "[198] . This is all the more true since the platform does not take into account any financial aid that may be provided by the local authorities199. "*Knowing how to obtain public subsidies, fiscal aid or effective intervention by the public authorities for the benefit of condominiums has become an essential management skill for the syndicate* "[200] since the absence of knowledge of these mechanisms deprives the syndicate of possible aid.

A co-ownership in financial difficulty due to a need for financing for heavy works must therefore approach the ANAH (and other organizations if necessary) to benefit from financial aid. Indeed, although they remain complex, the conditions for granting them are strict and they do not necessarily cover the entire amount of the work, they can accelerate the financial recovery. In addition to these subsidies, when a condominium in financial difficulty has a potential for densification, it can be interesting to mobilize this potential to find an additional source of income.

---

[194] Emilie PEYROTH, project manager at the private housing company in noisy le grand; CARASSO Jorge, "Sauver l'immeuble quand la situation se dégrade", *Le Figaro*, June 2015

[195] For further discussion see: www.anah.fr;

[196] Emilie Peyroth, supra.

[197] My project.anah.gouv.fr

[198] Delphine AGIER, project manager for the pact federation; CARASSO Jorge, "Saving the building when the situation deteriorates", *Le Figaro*, June 2015

[199] For example, the Programme d'accompagnement des copropriétés en difficulté (PACOD) in Metz, the Rennes old town center operation, Ecotravocopro... The amounts are generally less important than those of the ANAH, but the conditions are less strict ; In addition, certain measures facilitate the granting of ANAH subsidies (II.3.1)

200 MORELON Pierre, "Co-ownerships in difficulty: obtaining public aid for a syndicate of co-owners", *Loyers et Copropriété,* dossier 9, n°10, October 2015

### II.2.2 Mobilization of land potential as a source of potential income

The search for financial resources for the recovery of a condominium in financial difficulty may also consist of selling a portion of the condominium. In order to limit the inconvenience of a sale, the alienation of accessory rights is an effective device to generate cash flow and reduce expenses. However, there are a number of restrictions on this disposition.

In order to get out of its financial difficulties, a condominium must sometimes look for other potential sources of revenue, and section 3 of the law governing condominiums identifies several accessory rights which may be able to mobilize unused financial potential. The right to build and to raise are particularly interesting building rights in the context of the recovery of a co-ownership in difficulty. "*The right to build can be defined as the right to use the constructibility of a plot of land and to carry out any construction on it, in compliance with the legislative and regulatory rules* "[201]. In the context of the recovery of a co-ownership in difficulty, this right is however rarely exercised by the co-owners themselves. Since they do not have the necessary resources to invest, they generally turn to the alienation of this right202. Moreover, unlike the sale or donation of a part of the condominium, the alienation of the right to build does not deprive the condominium of any of its substance. On the contrary, it is intended to add one or more additional lots. This additional lot(s) must be taken into account in the new distribution of expenses and will allow for an economy of scale which will reduce the amount of the condominium owners' expenses. Moreover, if the syndicate sells the right to build, the capital gain is not taxable203 . The transfer of this right may be free204 or for a fee. In the case of a co-ownership in difficulty, it is of course preferable to transfer this right for consideration. Indeed, even if the transfer of the right free of charge may make it possible to reduce the charges, the main need of a co-ownership in.

---

[201] LELIEVRE Stéphane, "L'appartenance du droit de construire en copropriété", *AJDI*, pp.610-614, n°9, September 2011, paragraph 1

[202] LELIEVRE Stéphane, supra: most of the time the right to build is assigned to a third party or to a co-owner.

[203] Article 150 U-II-9 CGI

[204] BRANE Olivier, "Les nouvelles règles de la surélévation", *Droit et Patrimoine,* n°279, April 2018

*The equation is therefore sales price-construction cost=margin*

The challenge is to seek financing to bridge the cash flow gap. It is worth noting that all the ancillary decisions must be included in the construction cost.
"*In* the context of the alienation of the right to build, this device appears all the more interesting when it is the repair of the roof that has caused the financial difficulty. In the context of the alienation of the right to elevate, this device appears to be all the more interesting when it is roof repair work that has caused the financial difficulty. According to Giles FREMONT, director and president of the ANGC (Association Nationale des Gestionnaires de Copropriété): the elevation has become a source of financing for the syndicate that must be studied when the building requires major energy renovation work. If it turns out that the operation is promising to get out of the financial difficulty, the steps deserve to be engaged. However, in addition to the fact that it is not always technically possible to use these rights, these commitments will also have to respect numerous restrictions. These include compliance with town planning law. Indeed, despite the State's desire to densify the city, the various urban planning documents (SCOT, PLU, PPR...) can very quickly put a stop to the planned project. This fact is all the more true when there is a historical monument in the vicinity of the condominium in difficulty. As an example, in the context of an elevation, a building permit authorizing the elevation of a building was censured for its non-respect of "the diversity of heights [206]". In view of the difficulty of exercising this right, the legislator has come to relax certain rules.
*"Ordinance No. 2013-889 of October 3, 2013 on the development of housing construction, for example, has come to allow derogation from the rules of size, density, parking when issuing a building permit in the context of a right to overheight."* Similarly, *"Law No. 2014-1545 of December 20, 2014, on simplifying business life, made it possible to derogate from setback rules in the case of an over-elevation in a tense area. »[207]. For the right to build in general,*

---

[205] BRANE Olivier, supra

[206] CAA Paris, June 5, 2008, req n°06PA02679

[207] LEBATTEUX Agnès, "Les dispositions de l'ordonnance n° 2019-1101 du 30 octobre 2019 relatives aux modifications de la structure de la copropriété (syndicat secondaire, surélévation, scission) et aux copropriétés en difficulté", *Loyers et Copropriété*, n°2, février 2020; See also Article 1 Ordonnance n° 2013-889 du 3 octobre 2013 relative au développement de la construction de logement

*"It has given investigating departments the power to grant petitioners waivers to the provisions contained in urban planning documents. »*[208]. The trustee who wishes to initiate such a procedure must therefore approach the instructive services as soon as possible. Respect for the neighborhood can also quickly become an obstacle to the project. It will be necessary to estimate, for example, the loss of sunlight in order to avoid an action for abnormal neighbourhood disturbance. The respect of the co-ownership system itself can also be an obstacle for the mobilization of this right to build. Indeed, although the transfer of the right to build can bring about a recovery, the agreement of the majority of article 26, i.e. the qualified majority, is not always easy to obtain. Resistance to the project often arises as soon as significant investments are envisaged. In this case, for two-storey buildings without an elevator wishing to use their right to raise the height, the provisions of article R111-5 of the CCH209 may be a factor of opposition to the project. In addition, it is not always easy to "recover" certain private portions in order to build constructions.210 Faced with the reluctance of some, the BOUTIN law of 2009 relaxed the majority rule. The decision to transfer the right to elevate, which was initially covered by article 26, was lowered to an absolute majority in areas covered by a DPU.

In spite of numerous restrictions and constraints, the right to build can allow for the generation of cash flow and can be presented as a promising element for the financial recovery of a co-ownership in difficulty. If it is not possible to mobilize its land to generate cash, the co-ownership can always hope to mobilize the resources of private third parties.

### II.2.3 The use of private financing: a common benefit

In view of the large number of condominiums in difficulty, the legislator wished to encourage solvent third parties to undertake, in exchange for tax benefits, the financing of work to.

---

[208] GUIRRIEC Simon, *Les surélévations d'immeubles sous la double contrainte des droits de l'urbanisme et de la copropriété*, Master droit de l'Urbanisme Université de Bordeaux, p.106, 2016; See also Instruction du Gouvernement du 28 Mai 2014 relative au développement de la construction de logement par dérogation aux règles d'urbanisme et de la construction

[209] The installation of an elevator is mandatory in parts of multi-family dwellings with more than two floors.

[210] TGI Paris, 5 May 1981, a construction on a terrace implies the unanimous agreement; CA Versailles 3 June 1980, it is necessary the majority of article 26 to authorize a construction on a common part with private enjoyment; Cass. [3e] civ, 17 March 1980, refuses the right to build on a private terrace.

In the place of co-owners who do not have the means. Numerous measures have been put in place such as the Malraux program, the Denormandie measure, and the rehabilitation lease, but these have proven to be ineffective because the advantages are less and the risks are high.

The principle of the Malraux, Denormandie, Pinel program or the so-called "historic monument law" lies in the fact that those who undertake work on buildings requiring it benefit from a tax reduction. The Malraux program and the historic monument law concern real estate renovation programs undertaken by individuals on remarkable buildings211. The Denormandie and Pinel212 schemes concern energy renovation work in old buildings in specific municipalities. The tax benefits are not negligible since they can go up to 100% of the work for the "historic monument law". These devices therefore allow the cost of recovery to be borne by a solvent third party. On the other hand, according to the opinions of several investors, this is a system that will attract few people because it is of little interest to them213. Moreover, these schemes only concern very specific habitats, and most condominiums are not eligible for them. Aware of this failure, the legislator in 2014 completed these measures with a device that already existed to allow the recovery of individual houses: the rehabilitation lease214.

Faced with the impossibility of carrying out work due to a lack of cash, the rehabilitation lease appears to be an additional mechanism to facilitate the recovery of co-ownerships. Indeed, "*the contract allows them to entrust the renovation of their property to the lessee, then to occupy it once the work is completed as a tenant* "[215]. 215 In order to ensure that this operation allows for a sustainable recovery, even if the lessee assumes the management and costs of the unit, it is in the lessee's interest to let the lessor attend the AG so that he can take back his property in all conscience at the end of the lease.

---

[211] https://www.loi-malraux-monuments-historiques.fr/

[212] https://www.economie.gouv.fr/particuliers/reduction-impot-denormandie

213 As an example DARWIN Yann in his book "The profitable investor method", 2018: don't invest to pay less tax! Invest to make money.

[214] Article L. 252-1 et seq. of the Construction and Housing Code

[215] FAURE-ABBAD Marianne, "Bail à réhabilitation", *DALLOZrépertoire de droit immobilier*, Propriétaires occupants, July 2019, spec. n°3

In the same perspective, the taxes assumed by the lessee must allow the co-owners having resorted to this type of contract to be able, if possible, to save. These savings should be used to finance possible future works in order not to fall back into a spiral of degradation. This lease, which has a minimum duration of 12 years216 , can cover one or more condominium lots217 which has the advantage of being able to adapt to the situation. Indeed, in the case of work on a small part of the building, the disruption of the organization of the co-ownership will be less and more effective. And in case of works on the whole building, the organization will be able to have a total control on the rehabilitation project. The duration of the lease is variable according to the following criteria: the amount of the investment, the rental surfaces of the units to be rented, the levels of the rents practiced, the mobilizable subsidies. It is easy to understand that, for certain very heavy works, the duration of the lease is important. On the other hand, since the lease cannot be less than 12 years218 , it can be seen here that recourse to this type of contract should only be envisaged in the case of major works that cannot be assumed by the co-owners. "*At the end of the lease, the occupant (or his assigns) regains his status as co-owner by the simultaneous extinction of the real property right and the residential lease without compensation "*[219]. Although this contract has the advantage for the lessee of not having to buy the property, few contracts of this type have been issued. Between 1991 and 1995, only 333 rehabilitation leases were concluded.220 The main obstacle to this type of contract is that operations in condominiums in difficulty "*are costly and risky. It is therefore often a complex arrangement, which can last a very long time in the case of major works. This can also lead to a lack of confidence on the part of the co-owners with regard to the intervention, experienced as a form of guardianship "*[221]. The education of the syndic is therefore important. This can also be explained by the social purpose of this contract which limits its scope. Moreover, if the district concerned already occupies a part.

---

[216] Article L252-1 CCH

[217] SAINT-ALARY-HOUIN Corrine and MALINVAUD Philippe, "Bail à réhabilitation", *DALLOZ ACTION,*
file 130, The rehabilitation lease system, 2018-2019

[218] During the course of the contractual relationship, the parties may, at any time, decide to terminate it by mutual consent. In such a case, their will must be certain and unequivocal.

[219] FAURE-ABBAD Marianne, supra

[220] https://cours-de-droit.net/le-bail-a-rehabilitation-a121608952/

[221] COPROCOOP.fr, Portage de lots dans le cadre du Plan de sauvegarde du 12, rue Marcel Sembat à Montreuil, 2017, accessed 03/05/2020

If a significant amount of social housing is built, this is a real risk for the lessee. If no buyer comes forward, the operation will be loss-making. If it is not possible to conclude a rehabilitation lease and in the event of proven difficulties, a similar arrangement may be put in place. To do this, the judge222 must place the building under reinforced provisional administration223 which would make it possible to find pre-financing from an operator. However, one should not rely on this mechanism because, in view of the conditions to be met, this mechanism must "*remain exceptional*". [224]

We have seen that in the presence of bills that cannot be paid, it is necessary to approach public actors (ANAH, local authorities), private actors (HLM, investors), to study the possibility of mobilizing the land potential of the co-ownership, in order to hope to generate cash flow and to recover from this difficulty. As soon as all of these measures, including the provisional administrator, have not allowed the situation to be rectified in the long term, a serious deterioration sets in and risks causing the impoverishment of the district.

## II.3 Degraded co-ownerships: an urban consequence requiring a change in status

For legal practitioners, a co-ownership is considered to be in a state of disrepair when "*it presents risks for the safety of the co-owners and for which the conservation of the building seems complicated* "[225]. Faced with such a situation, the recovery226 allowing the return to a normal situation is long, difficult and sometimes doomed to failure. To prevent the advanced deterioration of the condominium from influencing the territory and impoverishing it, the community must accompany these condominiums, which requires considerable human and financial support (II.3.1) and sometimes requires considerable upheaval in the organization of the condominium (II.3.2)

---

[222] Upon a reasoned referral from the mayor, the president of the EPCI competent in housing matters, the prefect or the provisional administrator already appointed.
[223] Article 29-11 law of 1965
[224] Interview: Bernard CHEYSSON, lawyer at the Paris bar, *CCED n°6, 2016*
[225] ROUX Jean-Marc, Prévenir et redresser les copropriétés en difficulté, Edilaix, Point de droit, p.250, 2016
[226] Which requires a clean up of governance and a need for cash flow (see intro)

### II.3.1 Recovery achievable with community support

In order to facilitate the recovery of a deteriorated condominium, a community can set up a Programmed Operation for the Improvement of Housing (OPAH) and a safeguard plan227. These measures aimed at assisting co-owners in their recovery are effective, if implemented, since they allow co-owners to benefit from advantages that were previously inaccessible.

Created by a ministerial circular of June 1, 1977228 , the OPAH229 is a "*support mechanism for the recovery of one or more condominiums in terms of management, social and technical support* "[230]. The safeguard plan231 instituted by the law of November 14, 1996 is a similar support mechanism, but for specific condominiums experiencing particularly serious difficulties232.
"*The objective of these measures is to ensure and consolidate their maintenance in private status*. To achieve this, these mechanisms aim firstly to stimulate demand for ANAH subsidies by facilitating access to aid234. Unlike the *ad hoc* agent, the problem of unpaid thresholds is absent here. In spite of this flexibility, in order to be part of these mechanisms, the prefect235 for the Safeguard Plan and the mayor or the president of the EPCI competent in housing matters for the OPAH, must take the initiative, which is not always the case. "*This lack of action may, in part, call into question the interest of these measures, as a certain number of co-owners would like to see them implemented for fear of having to face more extreme procedures, such as the procedure for a state of*

---

[227] It can also set it up for co-ownerships in difficulty.

[228] In 1994 for the OPAH degraded co-ownership

[229] Article L 301-1 CCH

[230] Agence nationale de l'habitat and Ministère de la cohésion des territoires et des relations avec les collectivités territoriales, "Plan Initiative copropriétés Des dispositifs pour accompagner les interventions locales", 2019 [231] Article L 615-1 CCH

[232] "*One cannot help but make comparisons with the introduction of the recovery plan for a company in difficulty (art.L.621-54 C. commerce)*"; Daniel TOMASIN, "Plan de sauvegarde des copropriétés en difficulté", Report presented to the GRIDAUH (Groupement de recherche sur les institutions, le droit de l'aménagement de l'urbanisme et de l'habitat) place du Panthéon in Paris during the seminar on co-ownerships in difficulty, 2012

[233] Agence national de l'habitat and Ministère de la cohésion des territoires et des relations avec les collectivités territoriales, Plan Initiative copropriétés Des dispositifs pour accompagner les interventions locales, 2019,

[234] The usual delinquency threshold is no longer required

[235] Or at the request of the mayor, president of the competent Epci, residents' associations or a provisional administrator, (Article L 615-1 CCH)

*failure to act, which may lead to the expropriation of the building "*[236]. *It is* therefore in the interest of the syndic to approach these actors in order to benefit from these measures. The community should be made aware that the recovery of the condominium will also be beneficial for the interest of all. For example, faced with multiple uncorrected problems (poor maintenance, precarious residents, unsuitable co-ownership regulations) and a lack of preventive measures (no POPAC or VOC, significant unpaid debts, but below the threshold for the appointment of an agent), the Le François-West building complex in Vaulx-en-Velin fell victim to financial difficulties due to an urgent need for work237 . In order to recover, it demonstrated to the public authorities the difficulty of coping with the scale of the work. Subsequently, the public partners "*concluded that it was necessary to set up an OPAH for degraded co-ownership, with the aim of providing lasting solutions to the difficulties encountered* "[238]. After having defined an action plan, the operator sought financing to reduce the remaining costs for the co-owners as much as possible. "*The cost of the operation for the work alone was 3.3 million euros (80% of which was for energy saving work), i.e. an average cost per apartment of 41,800 euros. The public subsidies for each apartment financed an average of 35,000 euros, leaving a remaining cost of 6,800 euros per co-owner* "[239]. Alerting the public authorities to the situation is therefore an essential step in carrying out a project of this scale.

In addition to providing easier access to subsidies, these mechanisms ensure the efficient distribution and use of the aid obtained. Indeed, the OPAH agreement must specify the support and improvement actions to be carried out and the safeguard plan must elaborate a diagnosis of the situation and propose an action plan to resolve the difficulties. The Opah is a measure that is generally accepted by the representatives of the State in the field of urban planning, since 414.

---

[236] INFANTI Maeva, *Étude des difficultés juridiques inhérentes à la présence d'une copropriété dans ses relations avec les tiers,* Mémoire de Master Identification, Aménagement et Gestion du Foncier, ESGT, p.82, 2014

[237] HAINAUT Julie, " OPAH copropriété dégradée Une rénovation ambitieuse à Vaulx-en-Velin ",

*Informations Rapides de la Copropriété, CCED*, pp 7-8, n°4, March 2016

[238] HAINAUT Julie, above

[239] HAINAUT Julie, supra

In the same perspective, "*following the good results of this first operation, the community of Saint-Malo has decided to continue its actions by renewing this OPAH for a period of 5 years*. Concerning the safeguard plan, there are forty-eight in force to date. It is therefore a measure that is used, but only for serious cases.

"*The curative measures put in place by the public authorities are an opportunity for co-ownerships in difficulty*" [242] and deteriorated. In fact, they make it possible to respond to a need for cash flow and a potential need to improve governance. On the other hand, the needs of certain deteriorated condominiums are so great that it is impossible to straighten them out without considerably upsetting the organization of the condominium.

### II.3.2 Land portage as a last resort before the disappearance of the union

Faced with a very degraded condominium, *"it is not possible to resort to the common law of condominiums to durably rectify the situation "*[243] *as we have also seen previously.* Moreover, "*although the management of a condominium is a private matter, a serious deterioration of these buildings generates major safety and health problems* "[244]. Thus, the legislator has allowed the public authorities to considerably disrupt the organization of condominiums in order to facilitate their recovery, even if this may require the legal disappearance of the condominium.

In spite of the multiplication of measures aiming at financially supporting co-ownerships, urgent works requiring a significant amount are not always assumed by the syndicate.

---

240 ANAH, 2019 figure

241 ANAH AND SAINT MALO agglomeration, "Agreement for the implementation of an Operational Program for Prevention and Support of Condominiums (POPAC) in Saint-Malo Agglomeration", 2020-2022, p.3

242 https://arc-copro.fr/documentation/les-dispositifs-de-redressement-des-coproprietes-en-difficulte-pds-et- opah-which

243 Interview: Bernard CHEYSSON, lawyer at the Paris bar, *CCED n°6, 2016*

244 ROUX Jean-Marc, Prévenir et redresser les copropriétés en difficulté, Edilaix, Point de droit, p.250, 2016

This is why, faced with amounts that are impossible to recover, it is often necessary to force the defaulting co-owners to sell their property to third parties solvent245. In order to limit the number of disposals that are conducive to the arrival of slum landlords, the local authority has the possibility of derogating from the 1965 law by resorting to the partial expropriation of common areas246 , which would allow for a reduction in charges. If this measure is not sufficient to improve the financial situation of the condominium in the long term, the local authority must set up an Operation to Restore Damaged Condominiums (ORCOD)[247] . This operation consists, through an urban and social project, in treating the causes of the financial difficulty by resorting to a land holding operation248 which allows the community to take the place of the defaulting co-owners. This is a very good initiative since the recovery of a deteriorated co-ownership must be part of a global and coordinated strategy to be truly effective. The carrying of land completes the measures already undertaken in the fight to redress the financial situation of the co-ownership, but is however a long-term process249 . Indeed, in large condominiums, budgetary constraints and the absence of a desire to dispose of the property restrict the possibility of acquiring the units and rehabilitating them quickly. Although the ELAN law has facilitated the "unfolding" of an ORCOD250 , it is still a delicate operation. Indeed, *"feedback from the ORCOD demonstrates the inadequacy of the measures governing the rehousing of occupants*. Moreover, the operator can only buy back the properties at market price. The co-owners who still have a loan to pay will therefore not be able to sell their property if the amount does not cover their debts. In some municipalities, some condominiums have been declared ORCOD252 or even ORCOD of national interest,253 but this relatively effective mechanism is not used only on important co-ownerships.

---

[245] If all the co-owners are in default, they can all sell to a single buyer, but on the one hand this puts an end to the co-ownership regime and on the other hand, it is not always easy to find a buyer, especially for large property complexes.

[246] Article 615-10 CCH

[247] Art L 741-1 and 2 CCH ;

[248] This operation refers to the temporary purchase of lots by an organization that is responsible for restoring the building. HLM organizations, SEMs, SPLs and SPLAs are authorized to implement it [249] More than 15 years for the ORCOD-in Bas-Clichy

250 Removal of the condition of existence of a safeguard plan. Possibility of taking possession of degraded buildings to accelerate rehousing. Possibility of entrusting the piloting of an orcod to other operators than EPFs. Possibility for an EPF to benefit from the assistance of an EPA.

[251] GUILHEM Gil, "Copropriétés en difficulté - Garanties de recouvrement et syndicats de copropriétaires en difficulté", *JurisClasseur Copropriété*, Fasc.84, February 2018, spec. n°84

[252] Like Mantes-la-Jolie (more than 600 housing units), Argenteuil (more than 650 housing units)...

[253] Some condominiums are declared to be of national interest because their recovery requires exceptional resources. This is the case for the "Bas-Clichy" district or the "Grigny" district

All these difficulties preventing a rapid recovery of the co-ownership are not without danger, especially when the co-ownership is in such a state that emergency works cannot be undertaken for lack of means. This is why, in particularly dangerous co-ownerships, even if the state of deficiency254
In the case of a "*generally speaking, it is a severe measure with serious consequences* "[255], it will have to be pronounced. This state will put an end to the regime of co-ownership since the co-owners concerned will be expropriated256. The law of 1965 provides that the legal personality of the syndicate continues after expropriation for the purposes of the liquidation of the debts until the president of the judicial court puts an end to the mission of the provisional administrator257. This application for a declaration of deficiency is a last resort, while being an obligatory step for co-ownerships that cannot be rectified.

In view of the difficulty in restoring degraded condominiums, particularly large ones, the surveyor acting as a trustee must mobilize the public players as quickly as possible to facilitate the restoration, thus avoiding
to "reach" an advanced state of degradation.

Thanks to the numerous measures aimed at improving fragile co-ownerships, co-ownerships in difficulty and degraded co-ownerships, everything leads us to believe that the objective pursued by the legislator has been achieved. Nevertheless, it remains to be seen in practice if the recent measures mentioned so far are really understood and adopted by all (syndicates, co-owners, communities).

---

[254] Art. 615-1 paragraph IV CCH

[255] INFANTI Maeva, *Étude des difficultés juridiques inhérentes à la présence d'une copropriété dans ses relations avec les tiers,* Mémoire de Master Identification, Aménagement et Gestion du Foncier, ESGT, p.82, 2014

[256] It will then be advisable to accompany the expropriated person as best as possible. Overindebtedness procedure, rehousing in a social sector, renegotiation of the credit at the bank...

[257] Article 29-1 law of 1965

## Conclusion

The multiplication of measures aimed at reducing the number of condominiums with financial difficulties have certainly arrived late258 in French law, but they allow this objective to be met to a large extent.259 After having ensured that the drafter of the condominium by-laws has the tools to judiciously distribute the charges, thus preventing disputes which are conducive to the development of unpaid debts, the legislator has provided a framework for the mission of the trustee to ensure that he or she is no longer the cause of the advent of a financial difficulty. However, although there is still no clear data on this subject, the difference in supervision between volunteer and professional trustees, particularly with regard to the obligation of training, may slow down the effectiveness of the objective pursued. This strategy of fighting against the indirect causes of financial difficulties has also been extended to the co-owners themselves. Indeed, the implementation of a work fund and information tools allowing to anticipate the burden of future charges also aims at meeting this objective, but should be accessible to purchasers to improve its effectiveness. In addition, the recent provisions aimed at facilitating access to general meetings also facilitate the pursuit of the above-mentioned objective. Only the measures aimed at fighting against unauthorized co-owners show a modest effectiveness due to the fact that it is difficult to identify them. In order to accentuate the positive results of a good prevention, the community can, on the basis of indicators, intervene effectively with the co-ownerships showing signs of fragility in order to accompany them. If it is not possible to act in prevention, the surveyor acting as syndic will have specific tools according to his needs. In the context of a fragile co-ownership requiring a clean-up of governance, the recent provisions of the ELAN law will allow him to recover unpaid debts more quickly and effectively. If the co-owner is in default259 , the use of an ad hoc representative or a provisional administrator is an effective way to compensate for the co-owner's default, but the conditions for requesting their appointment are strict, thus reducing the effectiveness of these mechanisms. For a co-ownership in difficulty, requiring a need for cash, he will be able to count on public aids, in particular those of the ANAH, but he will have to master their complexity so that they are really effective. He could also mobilize the land to generate cash, but the numerous constraints, notably urban planning, make it an inefficient device. Finally, it could also rely on mechanisms that encourage private third parties to take the place of defaulting co-owners in exchange for benefits,260 but since the benefit is risky for the latter, these mechanisms remain ineffective. In the context of a deteriorated co-ownership requiring a heavy need for support, the measures that can be used by the local authorities (OPAH, PDS, ORCOD) are effective, but require patience before a normal situation can be restored.In view of the various successive reforms that have led to an increasing number of prevention and recovery measures for co-ownerships in financial difficulty, it seems interesting to consider the problem of access to co-ownership. Indeed, many difficulties could, it seems, be avoided if all the actors of the co-ownership are fully invested in the maintenance of their building and their living conditions within it.

---

[258] The first devices appeared in the 1990s

[259] Due to its own fault or to a lack of listening on the part of the co-owners

[260] such as the Denormandie scheme, the rehabilitation lease or others

## Bibliography


I- Works

Association des Responsables de Copropriété, " Savoir traiter les impayés en copropriété. Procedures, strategies, follow-up of the collection... all the means and advice to avoid the indebtedness of your co-ownership", *Vuibert*, 160 pages, 2011

Association des Responsables de Copropriété, " Les syndics bénévoles en copropriété ", *Vuibert*, 252 pages, 2015

L'union Sociale pour l'Habitat, "Constributing to the treatment of fragile and distressed condominiums," *Collection Cahiers*, 31 pages, 2017

ROUX Jean-Marc, "Prévenir et redresser les copropriétés en difficulté", *Edilaix*, Point de droit, 250, 2016

ROUX Jean-Marc and TOMASIN Daniel, " La Copropriété ", *DALLOZ Action*, 2018/2019

ROUX Jean-marc, " Les lots, éléments du patrimoine des propriétaires", *DALLOZ action*, 2018-2019

QUILICHINIPaule, "Compétencesdescollectivitésterritoriales :logement ", *Encyclopedia of Local Government*, DALLOZ, Chapter 15 (folio no. 4380), May 2018

SAINT-ALARY-HOUIN Corrine and MALINVAUD Philippe, "Droit de la construction, *DALLOZ ACTION*, 2018-2019

II-Legal Encyclopedias

BUCHER Charles-Édouard, "CO-OPERATIONS-Co-proprietors. - Duties of co-owners. "*JurisClasseur Civil Code,* Fasc. 31-1, January 2020

DERREZ Pascal, "Agence nationale de l'habitat. Subsidies for works: general conditions of attribution", *JurisClasseur Construction - Urbanism,* Fasc. 64-20, November 2014 (updated August 2018)

GUILHEM Gil, "Copropriétés en difficulté - Garanties de recouvrement et syndicats de copropriétaires en difficulté", *JurisClasseur Copropriété*, Fasc.84, February 2018

FAURE-ABBAD Marianne, "Bail à réhabilitation", *DALLOZ*, July 2019

LAFOND Jacques, "Scission de copropriété", *JurisClasseur Construction-Urbanisme,* Fasc. 94-50, July 2018

LAFOND Jacques, "Organization and operation of co-ownership", *JurisClasseur Notarial Formulaire,* Fasc. 60, September 2017

LAFOND Jacques, "Relations of the notary with the organs of the co-ownership", *JurisClasseur Notarial Form,* Fasc. 254, March 2015

LEBATTEUX Agnès, "Ordinance No. 2019-1101 of October 30, 2019 reforming the law of co-ownership of built-up areas", *JurisClasseur Copropriété*, Fasc.60-765, February 2020

ROUX Jean-Marc, "Synthèse-Droits et obligations des copropriétaires", *JurisClasseur Copropriété,* Fasc. 64, July 2019

VIGNERON Guy, "Common charges - Collection of charges - Procedures and guarantees of collection" *JurisClasseur Copropriété*, Fasc. 75-10, April 2010

VIGNERON Guy, "Secondary unions. Union des syndicats. Cooperative Union". *JurisClasseur Construction - Urbanism*, Fasc. 94-40, September 2011
VIGNERON Guy, "Synthèse - Administration de la copropriété", *Jurisclasseur Copropriété,* July 2019

IV- Scientific journals

- AXISA François, " La question des copropriétés en pré-difficulté ", *Droit et Ville,* n°73, pp19 to 25, 2012
BENASSE Christian, " La loi Alur n'apporte qu'une réponse partielle ", *La semaine Juridique Notariale et Immobilière*, n°20, mai 2014
BENILSI Stéphane, "L'indifférence du statut de la copropriété au caractère bénévole ou professionnel du syndic", *La Revue des Loyers*, n°990, October 2018
BOUYEURE Jean-Robert, "Le fonds de travaux", *Informations Rapide de la Copropriété,* pp.45-48, #628, May 2017
BRANE Olivier, "Les nouvelles règles de la surélévation", *Droit et Patrimoine,* n°279, April 2018
CAPOULADE Pierre, "Prévenir, repérer et traiter les syndicats en pré-difficulté", *AJDI,* March 2007, p.193
COHET-CORDEY Frédérique, " La cession d'un lot de copropriété : les pièges à éviter ", *AJDI*, p.813, 2012
- COUTANT-LAPALUS Christelle, "Some thoughts on civil liability and the syndic of co-ownership", *Loyers et Copropriété n°1,* January 2014, study 1, special issue n°9
DECHELETTE-TOLOT Pascaline, "Division into volumes: a facilitated exit for certain co-ownerships", *La Revue des Loyers,* n°948, June 2014
DEVRY Pascal, " Le point sur : The new ANAH aid for fragile co-ownerships", *cahier des copropriétés en difficulté,* pp.2, n°11, March 2017
- FREMEAUX Eliane, "The role of the notary in the life of the co-ownership", *Informations rapides de la copropriété,* n°582, 2012
GUITARD Pascal, "Co-ownership. - Responsibility of the syndic. - Infractions to the co-ownership rules. - Inaction. - Fault. - Mitigated responsibility of the voluntary trustee ", *AJDI*, p.798, 1995
GUEGAN-GELINET Laurence, "Loi ELAN et Copropriété", *La revue des Loyers,* n°933, January 2019
GUEGAN Laurence, "Loi ALUR : principales dispositions relatives à la copropriété", *La Revue des Loyers,* n°946, April 2014
GUEGEAN Laurence, " Différents motifs d'annulation de l'assemblée générale de copropriété ", *La Revue des Loyers,* n°938, June 2013
GUEGEAN Laurence, "Saisie immobilière et titre exécutoire", *La Revue des Loyers,* n°940, October 2013
HAINAUT Julie, " OPAH copropriété dégradée Une rénovation ambitieuse à Vaulx-en-Velin ", *Informations Rapides de la Copropriété, CCED*, pp 7-8, n°4, mars 2016
HAINAULT Julie, " Clichy-sous-bois : Launch of an ORCOD of national interest - a first ORCOD-IN ", *Informations Rapides de la Copropriété, CCED n°5, p.3,* 2016
HEUGAS Henri, "New social accession device: the social loan for lease-accession", *RDI,*

p.363, 2004
IVARS Camille, "The impact of the ordinance of October 30, 2019 on the procedural rules of co-ownership law", *Loyers et Copropriété,* n°2, February 2020, dossier 11, special issue n°19
LEBATTEUX Agnès, " L'amélioration des " modalités de gestion de la copropriété " par la loi Elan, *Loyers et Copropriété,* n°2, février 2019
LEBATTEUX Agnès, " Les dispositions de l'ordonnance n° 2019-1101 du 30 octobre 2019 relatives aux modifications de la structure de la copropriété (syndicat secondaire, surélévation, scission) et aux copropriétés en difficulté ", *Loyers et Copropriété,* n°2, février 2020
LEBATTEUX Agnès, "Étude sur les trois décrets parus en application des dispositions de la loi ÉLAN en matière de copropriété", *Loyers et Copropriété,* n°7-8, July 2019
LEBEL Christine, "Procedures applicable to syndicates of co-owners in difficulty", *Revue des procédures collectives,* n°5, September 2014
LELIEVRE Stéphane, " l'appartenance du droit de construire en copropriété ", *AJDI,* pp.610-614, n°9, September 2011
MONEGER Joël, "Co-ownerships: tradition and modernity", *Loyers et Copropriétés*, n°1, January 2020
MORELON Pierre, "Co-ownerships in difficulty: obtaining public aid for a syndicate of co-owners", *Loyers et Copropriété,* dossier 9, n°10, October 2015
PERINET-MARQUET Hugues, "L'immeuble face au dispositif duflot : changements et perspectives", *Droit et Patrimoine,* n°237, June 2014
- PERINET-MARQUET Hugues, "Access to housing and renovated urbanism Law ALUR of March 24, 2014", *La semaine juridique,* n°15, April 2014
PIPARD Dominique, "Economic and social legal issues related to degraded housing", *Le Lamy Collectivités territoriales,* n°7, November 2005
REGNAUT-MOUTIER, "Le traitement des difficultés des syndicats de copropriétaires", *Revue des procédures collectives*, n°3, May 2010
RODRIGUES David, "Le diagnostic technique global", *Informations Rapides de la Copropriété,* October 2018
ROUX Jean-Marc, " Réflexions sur l'assouplissement des conditions d'adoption des décisions en assemblée générale de copropriété ", *Loyers et copropriété,* n°6, June 2007
ROUX Jean-Marc, "The syndic and the "sleep merchants", *Informations Rapides de la Copropriété, CCED n°15,* 2018
SAINT ALARY HOUIN Corinne, " Les copropriétés en difficulté ", *Revue Droit et Ville*, n°52, p 103 et s., 2001
TOMASIN Daniel, "Responsibility of the co-ownership trustee", *AJDI*, pp.60,n°1, January 2002
TOMASIN Daniel, "Further reflections on co-ownerships in difficulty", *Droit et Ville, n°73, pp 11-18, 2012*
TORROLLION Jean-Marc, "Rôle et responsabilité du syndic dans la gestion des copropriétés dégradées", *Editions Francis Lefebvre*, March 2019
TURENNE Paul, "The voluntary syndic in co-ownership", *Informations Rapides de la Copropriété*, n°584, December 2012
VIEL Guillaume, CENAC Pierre and CETINER Devrim, "Co-ownerships in difficulty:

the contributions of the "Elan" law", *Droit et Patrimoine,* n°285, November 2018
VIGNERON Guy, " Recouvrement des créances du syndicat ", *Loyers et Copropriété,* pp.43, n°2, February 2016,

VI- University dissertations

BERGOZ Edouard, *Les scissions de copropriété comme outil de résolution de situations complexes,* Mémoire d'ingénieur Géomètre et Topographe, ESGT, 71 pages, 2017
BESCHI Sébastien, *Prévenir la dégradation des coprorpiétés récentes,* Master's thesis in Territorial Planning and Development, IUP Grenoble, 146 pages, 2008
CHAMOSSET Marie, *Accompanying condominiums in difficulty: prevention so as not to have to cure anymore?* Master's thesis for Urban Planning and Urban Project, Grenoble Institute of Urban Planning, 117 pages, 2015
DALBIN Hugo, *Does the application of the technique of division in volumes meet abusive uses?* Master's thesis Identification, Aménagement et Gestion du Foncier, ESGT, 60 pages, 2019
DALBIN Pauline, *Scission d'un grand ensemble initialement sous le régime de la copropriété en une division en volumes,* Mémoire de Master Identification, Aménagement et Gestion du Foncier, ESGT, 49 pages, 2015
DEBESSELLE Clémence, *La spécialisation des charges et des parties communes en copropriété,* Mémoire d'ingénieur Géomètre et Topographe, ESGT, 67 pages, 2016
DENIS Nicolas, *L'étude des techniques juridiques susceptibles de prévoir l'évolution du bâti en copropriété,* Mémoire d'ingénieur Géomètre et Topographe, ESGT, 57 pages, 2015
DUBOIS Diane, *L'utilisation des droits à construire au sein de la copropriété,* Mémoire de Master Identification, Aménagement et Gestion du Foncier, ESGT, 71 pages, 2015
FULCHERI Elisabeth, *Volume splitting, a solution for degraded condominiums?* Engineering thesis, ESGT, 102 pages, 2014
GAZEAU Anne, *Le statut de la copropriété dans la procédure d'expropriation*, Mémoire d'ingénieur Géomètre et Topographe, ESGT, 82 pages, 2016
GUIRRIEC Simon, *Les surelevations d'immeubles sous la double contrainte des droits de l'urbanisme et de la copropriété*, Mémoire de Master Droit de l'Urbanisme, Université de Bordeaux, 106 pages, 2018
INFANTI Maeva, *Étude des difficultés juridiques inhérentes à la présence d'une copropriété dans ses relations avec les tiers,* Mémoire de Master Identification, Aménagement et Gestion du Foncier, ESGT, 82 pages, 2014
JULLIAN Bastien, *Retour pratique de la loi ALUR sur les copropriétés dégradées et étude de l'une de ses innovations*, Mémoire de Master Identification, Aménagement et Gestion du Foncier, ESGT, 75 pages, 2017
PARISY Thomas, *La gestion des biens communs dans un ensemble immobilier :étude comparative entre les ASL et les unions de syndicats,* Mémoire de Master Identification, Aménagement et Gestion du Foncier, ESGT, 59 pages, 2019
SUSINI Lisa, *La surélévation des immeubles, un fort potentiel poussé par un contexte législatif favorable,* Engineering thesis, ESGT, 82 pages, 2019
VINCENT Sébastien, *L'intervention sur les copropriétés dégradées dans les programmes*

*de renouvellement urbain,* Master's thesis Urbanisme et Aménagement, Institut d'Urbanisme et d'Aménagement Régional Aix-Marseille, 100 pages,2017

III- Professional journals

AUCAME Caen Normandie, "Popac, A prevention device for weakened co-ownerships", *AUCAME,* pp1-4, n°83, April 2016

BADUEL Yves, "Fatality? Lack of technical mastery? ", *Geomètre, n°2170, p.28, June 2019*

-« The difficulties of voting by mail," *The CRA/UNARC Review,*
pp. 24-27, N°123, 1st quarter 2019

GUEGAN-GELINET Laurence, " Litigation for the collection of condominium charges after the ALUR and ELAN laws ", *Transversale immobilière,* pp.12-18, n°142, May-June 2019

V-General magazines

CARASSO Jorge, "Saving the building when the situation deteriorates," *Le Figaro*, June 2015.

BERTEAUX Alexandre, "Le recouvrement des charges impayées de copropriété", *Le Figaro*, March 2017

FRANCK Stéphanie, " Charges de copropriété impayées : quels recours *? ", Dossier Familial*, 2020

V- Institutional reports

Agence nationale de l'habitat, Les Opérations de Restauration Immobilière : Guide méthodologique, 2010, downloadable at:
https://www.anah.fr/fileadmin/anah/Mediatheque/Publications/Les_guides_methologiques/ORI_guide_methodologique.pdf

Agence national de l'habitat and Ministère de la cohésion des territoires et des relations avec les collectivités territoriales, Plan Initiative copropriétés Des dispositifs pour accompagner lesinterventionslocales, 2019, downloadable at:
https://www.anah.fr/fileadmin/anah/Mediatheque/Publications/Les_guides_methologiques / Recueil-outils-plan-Initiative-Coproprietes_WEB.pdf

Agence national de l'habitat, https://www.anah.fr/collectivite/traiter-les-coproprietes-fragiles-et-en-difficulte/choisir-un-outil-dintervention/, accessed 01/02/2020

National Housing Agency, https://www.anah.fr/fileadmin/forumhabitat/documents/78-20151121-Atelier_Redressement_judiciaire_des_coproprietes_Diaporama_20_novembre_2015-Support_de_presentation.pdf, accessed 31/01/2020

ANIL, Co-ownerships in difficulty / Ad hoc representative and temporary administrator, 2015

Consommation Logement Cadre de vie, Co-owners talk about their condominium, 2012

BRAYE Dominique, Preventing and curing condominium difficulties. A priority for housing policies, 2012
Directorate General of Planning, Housing and Nature, Loi ELAN Lutte contre les marchands de sommeil : les mesures, 2019
DORMOIS Rémi, " Le POPAC peut éviter de recourir à des opérations lourdes ", *Les Cahiers de l'Anah,* n°148, avril 2016
-ecologique-solidaire.gouv, Aides à la rénovation énergétique des logements : une nouvelle aide plus simple, plus juste et plus efficace, 2019
GELLY Ozone, "Un POPAC pour consolider les acquis et préparer le retour à un fonctionnement normal de la copropriété", *Les Cahiers de l'Anah,* n°148, April 2016, downloadable at:
https://www.anah.fr/fileadmin/anah/Mediatheque/Publications/Les_cahiers_Anah/cahiers-anah-148.pdf.pdf
Ministry of Territorial Cohesion and Relations with Territorial Communities, Aid to fragile or degraded co-ownerships, 2020
Ministry of Housing and Territorial Equality, Fighting substandard housing: The implementation of procedures in condominiums, 2014
TOMASIN Daniel, Report presented to the GRIDAUH (Research Group on Institutions, Urban Planning and Housing Law) at the Pantheon Square in Paris during the seminar on co-ownerships in difficulty, 2002
Ville-saint-malo, OPAH Co-ownership of degraded properties - Action Cœur de Ville de Saint-Malo, 2019
City of Metz, Bernadette condominium: land portage, DCM n°18-12-20-14, 2018, downloadable at:
https://metz.fr/pages/conseil_municipal/seances/cm181220/doc/70_d1545061273216.pdf

## VI- Legislative and regulatory texts

Codes:

Environmental Code
Urban planning code
Construction and housing code
Code of civil enforcement procedures
General tax code
General code of public property
Public Health Code

Laws:

Law No. 2018-1021 of November 23, 2018 on the evolution of housing, development and digital
Law n° 2016-925 of July 7, 2016 on the freedom of creation, architecture and heritage
Law n° 70-9 of January 2, 1970 regulating the conditions of exercise of activities relating to certain operations concerning real estate and business assets

Law n° 65-557 of July 10, 1965 fixing the statute of the joint ownership of the built buildings

Ordinance:
Ordinance No. 2019-1101 of October 30, 2019 reforming the law of co-ownership of built-up areas
Ordinance No. 2013-889 of October 3, 2013 on the development of housing construction

Decrees :
Decree No. 2019-502 of May 23, 2019 on the minimum list of dematerialized documents concerning co-ownership accessible on a secure online space
Decree No. 2018-11 of January 8, 2018 on the procedures for exercising the action for relief from foreclosure open to the creditors of a syndicate of co-owners in difficulty placed under provisional administration and making various amendments to the provisional administration procedure
Decree No. 2017-831 of May 5, 2017, on the organization and assistance of the National Housing Agency.
Decree No. 2016-1790 of December 19, 2016 on the declaration and prior authorization regimes for rental property
Decree No. 2016-285 of March 9, 2016 on the simplified procedure for the recovery of small claims
Decree No. 2015-764 of June 29, 2015, relating to the obligation of professional liability insurance for real estate sales agents
Decree No. 2013-610 of July 10, 2013 on the regulations for aid from the fund for the thermal renovation of private housing (FART)
Decree n°92-755 of July 31, 1992 instituting new rules relating to civil enforcement procedures
Decree n°72-678 of July 20, 1972 setting the conditions for the application of law n° 70-9 of January 2, 1970 regulating the conditions of exercise of activities relating to certain transactions concerning real estate and business assets
Decree n°67-223 of March 17, 1967 taken for the application of the law n° 65-557 of July 10, 1965 fixing the statute of the joint ownership of the built buildings. 1967

Circular:
Circular of February 8, 2019 on the strengthening and coordination of the fight against substandard housing

VII- Court decisions

**Judicial jurisdictions**
Court of Cassation:
Civ. 3e, Oct. 8, 2015, pourvoi n°14-19.245, *Dalloz actualité*, ROUQUET Yves, " L'action en recouvrement des charges incombe au seul syndic, 15/10/2015
Civ. 3e, January 22, 2014, n°12-29.368 : JurisData n°2014-000672
Civ. 3e, March 27, 2013, pourvoi n°12-13.012 : Administrer June 2013, p.54, obs J.R

BOUYEURE
Civ. 3e, January 24, 2012, n°10-27.944 : JurisData n°2012-000971, *Revue Loyers et Copropriété,* VIGNERON Guy, " Répartition des charges d'entretien et de conservation ",
Civ. 3rd, December 01, 2010, n°09-72.402
Civ. 3e, January 17, 2007, n°05-17.119 : JurisData n°2007-036928
Civ. 3e, April 26, 2006, n°05-11.986 : JurisData n°2006-033200
Civ. 3e, March 8, 2006, n°05-11042, published in Bulletin 2006 III n°599 p.49
Civ. 3e, May 11, 2000, n°98-17.026 : JurisData n°2000-001941
Civ. 3e, November 4, 1993, n°1679, pourvoi n°91-21.739
Civ. 3e, November 14, 1991, n°89-21.167 : JurisData n°1991-002840
Civ. 3e, Apr. 27, 1988, pourvoi n° 86-11.718 : Bull. civ. III, n° 80 ; Commented by LAFOND Jacques, " Organisation et fonctionnement de la copropriété ", *JurisClasseur Notarial Formulaire,* Fasc. 60, September 2017
Civ. 3e, May 13, 1987, pourvoi n°85-18.400
Civ. 3e, March 10, 1981, Appeal No. 79-
15.801 Court of Appeal:
Poitiers, 15 June 2011, JurisData n° 2011-031768 commented by VIGNERON Guy, "Common areas: exclusive right of use and allocation of charges", *Loyers et Copropriété,* n° 4, April 2012, comm. 126
Paris, 23rd ch. B, Nov. 25, 1999, commented by GIVERDON Claude, RDI 2000 p.89
Pau, September 4, 2006, JurisData n°2006-328864
Paris, January 11, 2001, JurisData n°1999-004298
Paris, September 29, 2005, JurisData n°2005-282555
Paris, March 27, 1996, JurisData n°1996-020631,
Versailles, March 23, 1989
Paris, March 4, 1986, commented by Inf. rap. copr. sept. 1986, p. 156

VIII-DALLOZ guidance sheets

- *Rehabilitation lease,* September 2019
*Global technical diagnosis,* July 2019
*Union*, July 2019 IX-
Maintenance

- Information meeting on the " Ecotravo Copro " device, Various actors of Rennes metropolises including Gilles DREUSLIN, Corinne LEGRAND and various trustees, on 15/01/2020
Alice LECAUDEY, Housing Officer, Saint-Malo Agglomeration, 19/05/2020
Corinne LEGRAND, referent in charge of technical urbanism at Rennes metropole, on 11/06/2020

X- Websites

ANIL.org, https://www.anil.org/immatriculation-coproprietes/, accessed on 12/12/2019
ARC-copro.fr, https://arc-copro.fr/documentation/1-apres-le-lancement-de-la-1ere-orcod-

de-france-retours-dexperience-de-larc-clichy, accessed 04/01/2020
Batinfo,https://batinfo.com/actualite/letat-a-verse-17-millions-sur-les-240-promis-pour-lutter-contre-lhabitat-indigne-a-marseille_14115?fbclid=IwAR0p7R-KEd2CA0jFAz3amAlrsf0oNWZ7YFp2Nk86bapE8oo6ihBx-ktULts,accessedon 05/11/2019
Batinfo, https://batinfo.com/actualite/la-reforme-des-coproprietes-validee-sans-sa-mesure-centrale-de-lobligation-de-plan-de-travaux_14101, accessed 02/11/2019
Batinfo, https://batinfo.com/actualite/nouveau-programme-dengagement-des-coproprietes-and-syndics-around-energy-savings_14238, accessed on 11/20/2019
Capital, https://www. capital.fr/immobilier/immobilier-le-psla-ce-dispositif-daide-a-lachat- sera-etendu-aux-logements-an-prochain-1355651?amp& twitter_impression=trueEnvoy,LOUKILAlexandre,viewed on 11/20/2019
Capital, https://www. capital.fr/immobilier/immobilier-le-dispositif-denormandie-pourrait-venir-plus-interessant-a-partir-de-2020-1355347, ASALI Sarah accessed 10/02/2020
Association des Responsables de Copropriété, https://arc-copro.fr/documentation/la-decheance-du-terme-version-loi-elan-une-super-procedure-de-recouvrement-des,accessed on 25/11/2019
PAP,https://www.pap.fr/patrimoine/copropriete/les-impayes-de-charges-de-copropriete/a4838, RAINFRAY R. and GALLOIS M., accessed 03/02/2020
Village-justice,https://www.village-justice.com/articles/super-procedure-recouvrement-loi-elan,30204.html, BOHBOT Charles and BOSCHER Paul, accessed 10/01/2020
Villagejustice,https://www.village-justice.com/articles/les-garanties-syndicat-des-coproprietaires-pour-recouvrement-des-charges,28831.html, BOHBOT Charles and JAMI Benjamin, accessed 10/03/2020
Weblex,https://www.weblex.fr/weblex-actualite/loi-elan-focus-sur-les-coproprietes, accessed on 15/01/2020
Weblex, https://www.weblex.fr/weblex-actualite/loi-elan-focus-sur-la-lutte-contre-lhabitat-indigne, accessed on 19/02/2020
Program-Malraux, https://programme-malraux.com/, accessed on 05/02/2020
Coproconseil,https://www.coproconseils.fr/recouvrement-simplifie-des-creances-loi-macron/, accessed 10/02/2020
Coproconseil,https://www.coproconseils.fr/coproprietaire-qui-ne-paie-pas-ses-charges-que-faire/, accessed 10/02/2020

## Table of appendices

**Annex 1[261]**
**The ANAH: an organization not to be overlooked**

[261] All the information in the appendices of this report comes from the ANAH website. https://www. anah.fr/

# L'ANAH AU SERVICE DES PARTICULIERS

**155 765 LOGEMENTS RÉNOVÉS**

+65 % par rapport à 2018

| 128 736 | 3 969 | 22 837 | 223 |
|---|---|---|---|
| logements de propriétaires occupants | logements locatifs | logements en copropriétés | logements pour des travaux d'office |

## TRAITEMENT DE L'HABITAT INDIGNE ET TRÈS DÉGRADÉ

**10 725** logements rénovés, **136,5** millions d'euros d'aides dont :

* Pour des travaux d'office

| → dont **4 522** logements Habiter Mieux | Propriétaires occupants | Propriétaires bailleurs | Copropriétaires | Communes* |
|---|---|---|---|---|
| Nombre de logements | 1 755 | 2 974 | 5 773 | 223 |
| Montants des aides (en M€) | 40,3 | 65,8 | 28,7 | 1,7 |
| Aide moyenne par logement (en €) | 22 965 | 22 113 | 4 971 | 7 950 |

En plus des aides aux travaux : **13,3 millions** d'euros pour financer 74 opérations de résorption de l'habitat insalubre (RHI).

## TRAITEMENT DE LA PRÉCARITÉ ÉNERGÉTIQUE

**116 995** logements rénovés, **760,5** millions d'euros d'aides dont :

| | Propriétaires occupants | Propriétaires bailleurs | Syndicats de copropriétaires |
|---|---|---|---|
| Nombre de logements | **109 359** | 3 421 | 4 215 |
| → Habiter Mieux sérénité | 40 895 | | |
| → Habiter Mieux agilité | 68 464 | | |
| Montants des aides (en M€) | 648,8 | 71,7 | 40,0 |
| Aide moyenne par logement (en €) | 5 933 | 20 945 | 9 478 |

## INTERVENTION SUR LES COPROPRIÉTÉS

**22 837** logements rénovés, **84,8** millions d'euros d'aides dont :

* Dont aides individuelles

| → dont **4 215** logements Habiter Mieux | Copropriétés en difficulté | Copropriétés fragiles | Accessibilité de l'immeuble |
|---|---|---|---|
| Nombre de logements | 19 467 | 2 686 | 684 |
| Montants des aides (en M€) | 72,7 | 11,9 | 0,2 |
| Aide moyenne par logement (en €) | 3 735 | 4 431 | 215 |

Les logements rénovés dans des copropriétés dégradées sont comptabilisés à la fois au titre du redressement des copropriétés et du traitement de l'habitat indigne et très dégradé.

## Appendix 2
## Aid remains complex, however

Aid to owner-occupiers:

► **Habiter Mieux sérénité** is an advisory service and a financial aid to help you in your project of global renovation of your home. Habiter Mieux sérénité concerns all the works allowing an energy gain of at least 25%. The financing is proportional to the amount of your work.

The amount of your Habiter Mieux serenity aid :
=>If you are in the "very modest resources" category:

- ✓ 50 % of the total amount of the work before tax. The Habiter Mieux sérénité aid is 10 000 € maximum.
- ✓ + the Habiter Mieux premium: 10% of the total amount of work excluding tax, up to a limit of €2,000.

=>If you are in the "modest resources" category:

- ✓ 35% of the total amount of the work before tax. The Habiter Mieux sérénité aid is 7 000 € maximum.
- ✓ + the Habiter Mieux premium: 10% of the total amount of work excluding tax, up to a limit of €1,600.

You can benefit from a larger bonus if your project meets the following three conditions: an F or G energy label before the work is done; work that allows for an overall energy improvement of at least 35% and a gain corresponding to at least a jump of two energy labels, your bonus is increased* under the following conditions
:

- ✓ 50% of the total amount of the work excluding taxes with a maximum aid of 15 000 for the "very modest" category and 35% of the total amount of the work excluding taxes with a maximum aid increased to 10 500 € for the "modest" category.
- ✓ + the Habiter Mieux premium raised to 20% of the total amount of work excluding tax, up to a limit of €4,000 for the "very modest" category and €2,000 for the "modest" category.

- **MaPrimeRénov** is the new State aid to finance your renovation projects "step by step". It concerns a list of works and equipment among the following categories: Insulation, Heating, Ventilation, Diagnosis and energy audit. The list of items that can be financed is defined according to the energy savings that each one allows to achieve. In order to obtain one of these two grants in 2020: You live in the housing you own; You must not exceed a certain level of resources; The work must be done by a company or a craftsman qualified RGE (recognized guarantor of the environment). This obligation only comes into effect on July 1, 2020 for Habiter Mieux sérénité. You have not received a PTZ (zero rate loan for home ownership) for 5 years (only in the case of Habiter Mieux sérénité); Your home is more than 15 years old or more than 2 years old.

The amount of your MaPrim'Rénov assistance:

The Anah grants you a lump sum for each item of work carried out. The lump sum is adjusted according to your level of resources and the energy gains made possible by the work done.

**Examples**
- You want to install an individual solar water heater? The aid is about 4000 €.
- You want to replace your gas boiler with an air/water heat pump? The aid is 7000 € *.
- You want to insulate the walls of your home (e.g. 100 m²) from the outside? The aid is 13 000 euros *.

4000 euros of which are energy saving certificates (CEE).

- **Habiter Sain or Habiter Serein**, work is necessary to make your home safe, comfortable and healthy. For example, it may involve installing or renovating water, electricity or gas systems. Or the installation of a bathroom and toilets. The reinforcement of the foundations or the replacement of a roof can also be concerned.

The amount of your Habiter Sain or Habiter Serein aid:

50% of the total amount of the work before tax. The Habiter sain aid is 10 000 € maximum, and the Habiter serein aid is 25 000 € maximum in the case of major works.

- **Habiter Mieux Copropriété**: new collective aid to finance energy renovation work in so-called "fragile" co-ownerships. To benefit from the Habiter Mieux

Copropriété aid, the condominium must: have been built before June 1, 2001, have at least 75% of residential lots occupied as a primary residence; be considered fragile, which means that its energy label is evaluated between D and G, and that its annual budget shows a rate of unpaid charges (between 8% and 15% of the total amount of the annual budget voted for condominiums with more than 200 lots; between 8% and 25% for condominiums with less than 200 lots). Under certain conditions, co-ownerships that are part of an operational program for the prevention and support of co-ownership (Popac) or a programmed housing improvement project (Opah) can also benefit from this aid.
This aid can be granted to the syndicate of co-owners for a program of work allowing a minimum energy gain of 35%.

The amount of your Habiter Mieux - Copropriété assistance:

- ✓ up to 180 € per dwelling for the cost of project management assistance (AMO);
- ✓ Financial assistance for the work, which can reach up to 25% of the total amount of the work excluding taxes. The Habiter Mieux co-ownership assistance is 5 250 € maximum per home including a bonus of 1 500 € for the energy gain.

► **Assistance for the renovation of common areas**: Thermal renovation work is necessary, such as the installation of a boiler in your building. Or heavy work on the structure of the building. It may also involve work to make the building or the apartments accessible to people who have difficulty getting around. Assistance can be paid directly to you. It can also be paid to your syndicate of co-owners if your co-ownership is in difficulty, or if it benefits from the Habiter Mieux - Copropriété aid.

- o Direct assistance to co-owners: You can apply individually or as part of a group application if there are other eligible co-owners. This unique file simplifies your steps. The procedure, the amount of aid and the conditions are the same as for the aid concerning your home. The aid is calculated on your share of the work. Please note that landlords who join your grouped application must sign a "Louer mieux" contract with Anah.
- o Assistance to the syndicate of co-owners: This assistance can be requested by your syndicate if your co-ownership is in great difficulty and is part of an Anah intervention scheme ("Programmed operation"). Or if your co-ownership is concerned by a situation of unworthy housing, a Plan de sauvegarde, or a provisional administration. Moreover, even if your co-ownership is not in difficulty, accessibility work can be financed by Anah via your co-owners' association. **50% of the total amount of the work (excluding taxes)** can be taken in charge. The maximum amount is 10 000 € per access.

| Cumulability of aid | MaPrimeRénov' | Habiter Mieux sérénité | Habiter Mieux copropriété | Habiter Sain/serein | Prime énergie | Aides locales |
|---|---|---|---|---|---|---|
| MaPrimeRénov | | X | X | X | √ | √ |
| Habiter Mieux sérénité | X | | X | X | √ | √ |
| HabiterMieuxCopropriété | X | X | | X | √ | √ |
| Habiter Sain/serein | X | X | X | | √ | √ |
| Prime énergie | X | X | X | X | | √ |
| Aides locales | √ | √ | √ | √ | √ | |

## LES CONDITIONS DE RESSOURCES

Vous pouvez bénéficier des aides de l'Anah si vos ressources sont inférieures à un plafond fixé nationalement.
Le taux d'aide de l'Anah peut varier selon que vous disposez de ressources "modestes" ou "très modestes".
**À partir de ce barème national, le contact local de l'Anah peut faire le choix des ménages prioritaires.**

Plafonds de ressources en Île-de-France *

| Nombre de personnes composant le ménage | Ménages aux ressources très modestes (€) | Ménages aux ressources modestes (€) |
|---|---|---|
| 1 | 20 593 | 25 068 |
| 2 | 30 225 | 36 792 |
| 3 | 36 297 | 44 188 |
| 4 | 42 381 | 51 597 |
| 5 | 48 488 | 59 026 |
| Par personne supplémentaire | + 6 096 | + 7 422 |

Plafonds de ressources pour les autres régions *

| Nombre de personnes composant le ménage | Ménages aux ressources très modestes (€) | Ménages aux ressources modestes (€) |
|---|---|---|
| 1 | 14 879 | 19 074 |
| 2 | 21 760 | 27 896 |
| 3 | 26 170 | 33 547 |
| 4 | 30 572 | 39 192 |
| 5 | 34 993 | 44 860 |
| Par personne supplémentaire | + 4 412 | + 5 651 |

**Are the prevention and recovery mechanisms available to co-owners in financial difficulty effective?**

Master's thesis C.N.A.M., "Identification, development and management of land" ESGT, LE MANS, 2020

---

**SUMMARY**

In order to reduce the large number of co-ownerships in financial difficulty, many measures have appeared since the 1990s aimed at preventing these situations. Initially based on the possibility of avoiding the causes of the arrival of a financial difficulty, other recent devices aiming at identifying and then accompanying the co-ownerships presenting signs of fragility have been created. In parallel to the prevention measures, the need for a heavier measure has led the legislator to build a framework which consists in providing support to the defaulting co-owners according to the level of their difficulty. These recovery measures are effective but must be used quickly in order not to reach a stage of no return.

**Keywords : Co-ownership, difficulty, Financial, Recovery, Prevention, Effective, syndic**

---

**SUMMARY**

In order to reduce the large number of co-ownership in financial difficulty, since the 1990s there have been numerous measures to prevent these situations. Initially based on the possibility of avoiding the causes of the arrival of a financial difficulty, other recent schemes to identify and then accompany co-ownership with signs of fragility have emerged. The latter are efficient but this requires involvement on the part of the trustee and the union in their co-ownership.In parallel with the prevention measures, the need for a heavier system has led the legislator to build a framework consisting of providing assistance to failed co-ownership, depending on the level of their difficulty. These righting devices are efficient but must be used quickly to avoid reaching a stage of non-return.

**Key words : Co-ownership, difficult, financial, turnaround, prevention, efficient, managing agent**

Printed by Books on Demand GmbH, Norderstedt / Germany